The Unconscious in Neuroscience and Psychoanalysis

The Unconscious in Neuroscience and Psychoanalysis presents a unique and provocative approach to the assimilation of these two disciplines while offering a thorough assessment of the unconscious from a neuropsychoanalytic and Lacanian perspective.

Marco Máximo Balzarini offers a comprehensive overview of Freud's theory of the unconscious and its importance within psychoanalysis, before looking to how it has been integrated into contemporary neuropsychoanalytic work. Paying close attention to the field-defining work of neuropsychoanalysts such as Mark Solms, Francois Ansermet, and Pierre Magistretti, Balzarini considers the dichotomy between neuroscience and psychoanalysis, and the omnipresent debate on if and how they should be integrated when working with the unconscious. Throughout, he provides a fascinating Lacanian interpretation, showing how the work of Lacan can offer a new way of developing the dialogue and understanding around this vital topic.

Part of the Routledge Neuropsychoanalysis Series, this book will be of interest to any psychoanalyst seeking to explore the foundations of the relationship between neuropsychoanalytic and Lacanian ideas in their clinical and theoretical work.

Marco Máximo Balzarini is a practitioner of Lacanian-oriented psychoanalysis in the city of Córdoba, Argentina, and is a Professor of Psychoanalysis at National University of Córdoba (UNC). He has published widely in national and international academic journals, books and newspapers.

The Routledge Neuropsychoanalysis Series
Series editor: Mark Solms

The attempt to integrate the findings and methods of psychoanalysis with those of the neurological sciences can be said to have begun in 1895, with Freud's Project for a Scientific Psychology. Ongoing, sporadic efforts continued throughout the 20th century. However, the field really took off when the journal Neuropsychoanalysis was founded in 1999 and the International Neuropsychoanalysis Society was established in 2000. Ever since, a themed annual congress has been held in different cities around the world. Today, it is fair to say that these efforts have generated the most rapidly growing and influential body of knowledge and clinical practice in the broader field of psychoanalysis.

The establishment of this book series in 2023 marked another important milestone in the development of the field. Under the editorship of Mark Solms, the co-chair of the International Neuropsychoanalysis Society, it publishes books by leading proponents - and critics - of neuropsychoanalysis. The books in this series focus not only on the scientific findings of neuropsychoanalysis and on its theoretical yield, but also on its history, its philosophical implications and its clinical practice, as well as its ramifications for neighbouring disciplines and for the mental and neurological sciences as a whole.

The Spirit of the Drive in Neuropsychoanalysis
Mark Kinet

The Unconscious in Neuroscience and Psychoanalysis
On Lacan and Freud
Marco Máximo Balzarini

For more information about this series, please visit: www.routledge.com

The Unconscious in Neuroscience and Psychoanalysis

On Lacan and Freud

Marco Máximo Balzarini

Routledge
Taylor & Francis Group

LONDON AND NEW YORK

Designed cover image: timandtim / Getty Images

First published in English 2024
by Routledge
4 Park Square, Milton Park, Abingdon, Oxon OX14 4RN

and by Routledge
605 Third Avenue, New York, NY 10158

Routledge is an imprint of the Taylor & Francis Group, an informa business

Published in Spanish by Grama Ediciones 2023.

British Library Cataloguing-in-Publication Data
A catalogue record for this book is available from the British Library

ISBN: 978-1-032-60286-8 (hbk)
ISBN: 978-1-032-60284-4 (pbk)
ISBN: 978-1-003-45847-0 (ebk)

DOI: 10.4324/9781003458470

Typeset in Times New Roman
by codeMantra

Contents

Acknowledgments   *vii*

Preface   *ix*

Introduction   1

1   The concept of the unconscious in current neurosciences   3
Neuroplasticity and memory 3
Procedural reality 9
Internal and autonomous reality 11
The mental unconscious 14
Between two neurobiological theories 19
Drive, jouissance, and motor activation 25
Life outside of the being that speaks 29
Cognitive machine 33
The id, affective consciousness 39
The neurochemical basis of libido 40
There is no intracerebral relationship 43
RSI brain mapping 44
Psychoanalysis will disappear if it is not saved by neurobiology 45
Interview with Dr. Carlos Daniel Mías 50

**2   The intention in Freud's saying: There is no relationship
       between the unconscious and the brain**   64
Beginnings of the relationship with science 64
Soul surgery 67
*From a conception that starts from the brain to another that
       excludes it 69*
Difficulties in the project 70
The last man 73
Drive fixation does not take place in the brain 75
Of a non-biological body 76
To make a mistake is to say it right 77

Scientific infatuation 79
The problem of scientific induction 85
Senses of the term unconscious 86
The term "Io" 90
The turn of the 1920s 91
Beyond scientific experimentation 94
Logic of the unconscious in Freud's work 96

3   **The non-relationship of the unconscious and the brain in
     Lacan's reading of Freud**   102
The neuroscientific unconscious and ours 102
The contemporary doctor 103
Brain, therefore I am 104
Body that enjoys 104

4   **Consequences of the biologization of the unconscious**   107
From a paradox within these neurosciences 107
Neuroreligion 108
Imaginization of the unconscious 110
The real in neuroscience and psychoanalysis 113
The neuropharmacological model 114
The right to health 116
Neuroscientificism 118
Science fiction 120
The ascent of the signifier neuro 122
Psychoanalysis position 123
Revolution or subversion 124
Desire to wake up 125

5   **Neuroscientific imputations to psychoanalysis**   131
Therapeutic ineffectiveness 131
No falsifiability 133
It's not a science 134
Diffusion deprivation 135
Universe of discourse 137
Insufficiency in the explanation of the mind 138
Postpone relief 138
It does not submit its operation to expert evaluation 138
The non-existence of the sage 139
It is not welcome! 139

Final thoughts   142

Bibliography   *145*
Index   *159*

Acknowledgments

To the memory of Néstor Efraín Cornejo, who, with his intellectual background and sensitivity after the harshness of the police officer, knew how to accompany my first years of life;

to Dafne Sánchez, for her love;

to Lili Aguilar, for hosting the difference that I didn't know lived in me and accompanying me, each time, to find where it was;

to Ale Rostagnotto, for his direction, his support, guidance, and contributions to this research;

to Mariana Gómez, for her generous preface, for guiding the transfer of work that links me to psychoanalysis and for her contributions at the beginning of this research;

to Zoe Meyer, for bringing me the bibliographical suggestions resulting from the evaluation process towards the publication of this research, readings that I did not know and that improved it;

to Mark Solms, for welcoming this proposal and deciding to publish it in the section he directs.

Preface

Mariana Gómez

The Unconscious in Neuroscience and Psychoanalysis, the title of this book is sufficiently provocative. At the same time, it is challenging for its author Marco Balzarini. I have known him since he was very young and I know his taste for challenges, but I also know his permanent restlessness and desire to investigate and write. Therefore, I was not surprised to read this work. But I have learned a lot from it.

With a rigorous and precise style, the author delves into the usefulness of the concept of the unconscious, and its productions, in social and scientific discourse. But, at the same time, he analyzes how from neuroscience, and according to certain references, we find that this concept aims to be explained from neuronal functioning. Despite the fact that, as anticipated in its introduction, there is an offensive by the neurosciences toward psychoanalytic discourse. What is it that bothers, touches, psychoanalysis in some scientists?

Thus, the book presents the attempt of some neuroscientists to articulate both theories and make this epistemic intersection a relevant clinical utility. In my opinion with little or no success.

Despite this, Marco proposes in this book to analyze the legitimacy of the positions that support the unconscious–brain combination. In this sense, the question that guides this text is, how does the unconscious of Freud and Lacan differ from the unconscious conceived by current neurosciences? The definition of mental health, in my view, could introduce us to this question.

The World Health Organization (OMS) defines mental health as a state of well-being in which the individual is aware of his or her own abilities, can cope with the normal stresses of life productively, and is able to make a contribution to his or her own community. Mental health is thus defined by this organization in its positive dimension in that it considers health in general terms as "complete physical, mental, and social well-being and not only the absence of illnesses or diseases".

Likewise, the OMS establishes that there is no official definition of what mental health is and that any definition in this regard is determined by the social, the cultural, the subjective, and the discussions between the various professional theories.

The OMS thus encourages the use of psychotherapies distributed among various orientations: psychoanalytic, cognitive-behavioral, systemic, among others.

In this way, the *Diagnostic and Statistical Manual of Mental Disorders of the American Psychiatric Association*, in its successive editions, proposes a countless number of classifications and disorders that are built on the basis of an idea of normality, insofar as some of this has been altered, considering also that it is possible to restore said normality by suppressing the disorder. This would imply a way of proceeding that is the same for everyone, supported by an idea of healing linked to re-establishing the state of harmony prior to the disorder. In this way, the state of mental health would be linked to that of balance.

For his part, and from the psychoanalytic perspective, Freud will maintain in his work *Culture and its Discontents* that happiness and well-being are only achieved momentarily, as the subject encounters his symptom again and again. Therefore, the search for happiness and the symptom constitute an unavoidable tension. The symptom reminds us that homeostasis is not what reigns in the human world. Thus, the symptom is what breaks the illusion of complete mental health.

With Freudian and also Lacanian theory, we know that although the practice of psychoanalysis entails therapeutic effects, these are not what guide it. Hence, the term "therapeutic efficacy" is not appropriate to the field of psychoanalysis, if by doing so we hope to obtain the absolute solution to suffering. The work of the psychoanalyst, on the contrary, does not seek to "cure" or do good, in the manner of a *cure fury*. For this reason, Lacan, in his *Seminar 7*, will present the analyst's desire as a desire not to cure, while directing the treatment toward cure would mean "wanting the good" of the subject. That is not the ethics by which psychoanalysis is guided. The ethics of psychoanalysis is based on the development of a knowledge about the symptom itself, where the cure will come in addition.

In this framework, psychoanalysis, which can only exist in a Rule of Law, is a discourse and a way of approaching subjectivity in its singularity, with a method and an episteme that are its own. From this perspective, it constitutes a practice from which we will no longer speak of illness or disorder, but rather of subjective suffering. A condition that the subject will have to reduce, according to his own possible solution.

Therefore, the author's intention to analyze the tensions between two disciplinary fields whose object is the human subject is extremely interesting. Tensions that Marco will know how to resolve in his own way, in the idea that underlies these pages, and that is that neuroscience proposes a subjectivity that involves "for all", but that there is something that exceeds that uniformity and that is "singular suffering", impossible to classify and standardize. This postulate is what will lead readers to verify that psychoanalysis goes against the grain of the logic of goodness and happiness, but rather, goes along the path of singular functioning.

Positioned, then, in the discourse of psychoanalysis, we will not consider mental health as a possible and interchangeable good in the health market, with the ideal of a cure. But as the impossibility of achieving total well-being in a job of subjective involvement and access to know-how as a product of an analytical journey, outside the ideals of speed and efficiency.

From there, and assuming the distinction is necessary between what psychoanalysis proposes as a cure, on the one hand, and what other neuroscientific practices understand as mental health, on the other, Marco Balzarini's question about the difference between the unconscious of Freud and Lacan and the – forced – unconscious that current neurosciences conceive, in my reading, is answered in this text that the author offers us. And, as Lacan said, whoever has a question has his answer.

And what I capture in this book is that although Marco already had his answer, he nevertheless took all the trouble to answer it in detail. This is what the reader will find in these pages: exhaustiveness, reflection, and its own enunciated position.

September 2023

Introduction

Neurosciences are very present in common discourse. News of new discoveries about perception, memory, sleep, etc. appear in most of the world's newspapers. They are also in TV series, in councils, and in State policies, while psychoanalysis is not there, why? Due to the reserved style of psychoanalysts, which enables other voices to speak, among them certain voices of science, which today are trying to demonstrate, in a large number of cases, the determining influence of neuronal bases on the psychic phenomenon. It is not new that a discipline has credibility if it supports the scientific rigor of empirical demonstration; what is new is that researchers who call themselves psychoanalysts seem to love it. Indeed, great neuroscientists are affirming that the Freudian concept of the unconscious, as well as Freudian metapsychology, the concept of jouissance, and a set of concepts from Lacan's teaching, can be explained with neuronal rigor. Furthermore, they claim that Freud's intention was to reduce psychoanalysis to biology. In other words, the old tradition of searching for the dimension of existence in the observable record of the organ is being updated under the aegis of neurobiology.

Analysts must not remain silent. The community of psychoanalysts needs analysts with sufficient training in the Lacanian School to pierce the discourse of the master, supported by neuroscientists with a lot of press, such as the case of Facundo Manes in Argentina, and to pierce the escalation of neurosciences, especially now that cognitivism has reached Parliament. One day we are going to get up and psychoanalysis is going to be prohibited. At this point, we return to the question that Jacques-Alain Miller formulates: "what is psychoanalysis to deserve this offensive and [...] what is psychoanalysis to stop it, and to appear at least today, for the moment, as a nucleus of resistance to it?" (2004c, p. 155).

Indeed, the discussion is productive, but not when it comes to a critique between doctrines. For this reason, we do not intend to question the specificity of neurodiscourse, but to analyze the legitimacy of the positions that support the unconscious-brain combination. In this sense, the question that guides this text is: how does the unconscious of Freud and Lacan differ from the unconscious that current neurosciences conceive? This question contains a statement: there are differences. This is the hypothesis we are going to test. Psychoanalysis and neurosciences are, as Kuhn (1964) indicates, two disciplinary matrixes that cannot

DOI: 10.4324/9781003458470-1

be measured, that do not represent progress, cannot be compared, they are two "incommensurable ways of seeing the world and of practicing science in it" (p. 25) and "proponents of incommensurable theories cannot communicate with each other, at all" (p. 303).

It is true that psychoanalysis was born under the postulates of neuroscience, but that combination not only was not enough for Freud to study clinical phenomena, but what Freud meant was that there is no such combination. Saying does not mean that Freud said what he wanted to say, that was not the objective of his project, but what he said in a sustained manner, what he demanded, or what he said between the lines. Let's revitalize his saying as the integration movement has been renewed and is advancing. We are going to support Freud in a way that without doubt. The doubt comes from the training of those who support it. It is a question of seeing in whose hands psychoanalysis will be left: whether in the great neuroscientists or in the Lacanian-oriented practitioners of psychoanalysis.

Reference

Miller, J.-A. (2004c). Verdad, probabilidad estadística, lo real. In *Revista Lacaniana. Las prácticas de la escucha y sus argumentos* (2). Buenos Aires: EOL.

Chapter 1

The concept of the unconscious in current neurosciences

Neuroplasticity and memory

Recent research in neurosciences has taken a decisive turn from the concept of neuroplasticity. Neuroplasticity means that the brain changes based on individual experience, that the brain is naturally (i.e., genetically) programmed to be open to unnatural (i.e., not genetically programmed) influences. Epigenetics, systems that control the extent of gene expression, extend this logic beyond ontogeny. Environmental experiences can be encoded into epigenetic effects on the individual and even on the individual's offspring. The social world can thus distort the genetic patterns of the brain, unthinkable without the environmental histories that shape it. In this way, epigenetics reveals a genetic indeterminism that goes against the determination of programming by nature to reprogramming through breeding. The key tool for this is neuroplasticity, one of the main channels through which the endogenous body is denatured beyond the biological (Johnston, 2013; Dall'Aglio, 2020b; Ansermet and Magistretti, 2006).

Françoise Ansermet, a French psychoanalyst and member of the World Association of Psychoanalysis, and Pierre Magistretti, an Italian neuroscientist of world importance, emphasize that neuroplasticity opens a "gap" between the brain, subjectivity, and its environment. They attack the idea that we receive genetics and meaning already given by stating that the brain is open. By which they say that they come close to the Lacanian idea that we are not born with a body. Likewise, John Dall'Aglio, from Duquesne University in the United States, affirms: "the organism as an organized unit does not exist from birth" (2020b, p. 717). With this perspective, the authors say that they have recovered the value of the social Other and hope to alleviate the anti-reductionist objections coming from psychoanalysis.

In this sense, various authors (Johnston, 2013; Ansermet and Magistretti, 2010) have studied the cerebral insula in monkeys and humans. They have found that this brain structure is a kind of filter that explains the difference between animal instinct and human drive. A kind of filter that means that the human being does not react as a reflex that tends toward homeostasis, but that the regulation process is mentalized and free of automatisms typical of lower species. Thanks to insular mediation, in terms of neuroanatomy and neurophysiology, the unreflexive and

DOI: 10.4324/9781003458470-2

automatic regulation of the vital functions of life by the brainstem is interrupted. Insular mediation remains in relation to mentalization and cognitive representation to prevent vital functions from being reduced solely to instincts. Thus, they are directing organic homeostasis through matrixes of imaginary representations inscribed in the insula on the sensory and perceptual phenomena that are becoming signifiers. In turn, the insula has the particularity of having a remarkable plasticity that allows both the retention and the reworking of its representational contents. If the insula is in charge of processing stimuli and solving internal–external representations, we have the place where instincts are trapped and processed by cognition, that is, we have an island in a sea of instincts. With this, Ansermet and Magistretti (2010) say that they are in line with Lacan by highlighting the mediating role of language beyond biology, demonstrating a biology that is not biology, the genetically determined that is not genetically determined, because if the puppy human is a clean slate, a blank sheet capable of being written and molded, the genetic determination has no foundation here. So, neuroplasticity comes to oppose the genetic thesis under this conception of the insular cortex.

Ansermet and Magistretti's (2002, 2006, 2010) thesis is that the autonomy of the psychic subject results from the fact that his plastic brain is individuated to the point of being completely unique and that such singularization, facilitated by plasticity, is equivalent to freedom. The subject is the exception to the universal, but with neurobiological foundations. This allows them to say that, reading Lacan, they have gone beyond biology because they have identified an area exclusive to human beings, distinguishing them from their closest primate relatives, which distances human beings from instinct and places them beyond biology. It is a viable quasi-naturalist theory of the autonomous subject, that is, a Lacanian neuropsychoanalytic biology of freedom.

For his part, Eric Kandel, an American neuroscientist, received a Nobel Prize in 2000 for having contributed to neurophysiological studies by proving the relationship between memory and synaptic neuroplasticity in snails and mice (Kandel, 2001b, 2009a). Thus, Kandel was able to establish certain principles of the mind, among which the plastic quality of neurons modifies the brain. If the brain is genetically indeterminate, it is not a rigid, preformed unit, "it can change its property depending on its state of activity" (Changeux, cit. Castanet, 2023, p. 62). "The neuron is no longer apprehended in its cellular uniqueness, but in its connectivity with other neurons" (Castanet, 2023, p. 63). This is about the description of a brain that can be shaped by new synaptic connections, for example, we can think of children who go to school and who come home every day having learned new things, that is, they are no longer the same.

From this perspective, the memory trace is the material inscription of the experience "inside" the brain. That is, the cerebral structure is the intimate reflection of the experience. The dilemma between nature and culture would be overcome because the materiality of the mental image cannot be questioned (Castanet, 2023). One is born with an immature brain, endowed with great plasticity, new traces are created, and nothing is fixed in the neurons (Dehaene, 2015), which makes it possible to modify not only the traumatic reference, but also, and this is another of the principles established by Kandel, genetics.

We owe Kandel the studies on Aplysia, a species of sea hare that has 20,000 neurons while humans have billions (Yellati, 2021; Castanet, 2023). He discovered that the memory imprint is embedded in a new neuron. Due to the fact that this experience was verified based on a new neuron, he concluded that the changes remain inscribed forever. Therefore, the brain changes, due to the environment, learning, the context, psychotherapy, everything affects and remains inscribed in the brain. Any lived event is marked instantly and can persist as a kind of incarnation of time.

Ansermet and Magistretti (2006) have worked in this direction to demonstrate the link between experience and neuronal plasticity, defining synaptic footprint as a representation of a specific experience of the external world. They maintain that the brain remains open to experience, and that this experience, every day, modifies the brain and also the unconscious. His thesis is that the brain can be modified from any experience, which demonstrates the neurophysiology of the unconscious.

The idea shared by all these authors is that the traces left by experience are not static. That is to say, any experience leaves a trace and this continues to change based on additional experiences that are divorced from the original trace, producing a gap that they designate as the real in the brain. They emphasize the neuroplasticity of imprints, imprints constantly reassociating, strengthening, or weakening according to the principles of synaptic memory. They don't say that the imprints are true to the experience, they say that the original imprint of the experience changes due to neuroplasticity. Plasticity then assumes that the experience is inscribed in the neural network (Ansermet and Magistretti, 2006; Kandel, Schwartz, and Jessel, 2001).

Now, if the subject registered and classified all of reality, literally all of it, it would be a hard drive with unlimited storage. The unconscious has nothing to do with it. What Freud calls the unconscious is the trace that was not inscribed; it is what has been erased from the trace that left some experience of satisfaction, which is irretrievably lost. "The trace of perception is inscribed nowhere. The unconscious is precisely the lack of this trace, not the trace of something that happened" (Bassols, 2011a, p. 98, 2011b). That is why speaking has nothing to do with memory because by speaking one creates the language (Miller, 2014b).

In the story "Funes, the memorioso" [Funes, the memorious] Borges describes the suffering of remembering everything or of not being able to forget. Funes is like a machine keyboard with the delete key removed. "It is a subject that keeps in his memory everything he perceives, everything he thinks and experiences from birth, with the terrible problem that he cannot forget anything" (Bassols, 2011a, pp. 98–99). Borges thus describes the torture of not being able to forget, where memory and perception are not excluded, where the signifier has not operated, and that is why it is torture. Finally, Funes asks to forget, "he asks for a bit of the unconscious, he asks that language allow him to separate himself from reality" (p. 99). Memory implies a subject. Traces are not recovered, they are made by a subject. The difference between perception and memory that Freud wrote in his first letters supposes a subject. Locating the unconscious in memory, accessible by plasticity, is to reject it as if it were no longer in becoming (Cuñat, 2019).

With this notion of neuroplasticity and its link with memory, current neurosciences have gone beyond the genetic model (fixed program) toward the model of plasticity (modification). Already in the 19th century, due to Darwin's intellectual action, it was accepted that the being that adapts to the environment, that is the being that produces changes in its biology, would survive. Natural selection verifies that the genotype of a living being evolves from experience with the environment. The new thing that these neurosciences propose is that this theory is now supported by Freud. In this sense, Ansermet and Magistretti (2006) state that Freud understood the role of this plasticity in learning mechanisms and that it took mnemic traces as a correlate of perception. In "Letter 52" they read that between the two extremes, perception and consciousness, there is a series of successive transcriptions that constitute the unconscious. Said transcriptions, the authors maintain, are carried out by the mechanisms of synaptic plasticity in the manner of an image that is linked to another by the information it receives from the senses. Such neural chaining reactivates memory.

It is not about the wandering rigor that Lacan gives to the chaining through the signifier, but about an unequivocal capture in networks of neurons, coagulated one into another, that reactivates, without loss, the memory image. It is a positive trace, objectively observable, while for Lacan (2009f, 2007) the trace is negative since it refers to its absence, the signifier being the sign of that absence; the more one tries to erase that trace, the more the trace insists as a signifier. However, for Ansermet and Magistretti (2006), the unconscious is in correspondence with sensory perception; traces, signs, and representations would be equivalent concepts:

In this way, when the brain perceives and registers in the form of a trace the stimulations coming from the outside world, which lead to the construction of a psychic trace (transcription of an external reality), then there can be a correspondence between the trace (signifier) and external reality (signified): the signifier corresponds to the signified. This correspondence, which is of a conscious nature and which reveals cognitive processes, constitutes the basis that allows us to locate ourselves at different points in reality.

(p. 94)

The signifier–signified bi-univocal relationship implies that there is only one name for each thing and only one thing for each name. On the contrary, for Lacan the signifier allows to name and to name is to create. The use we make of the signifier is to deceive about what has to be signified, because the signified is always missing. While for Ansermet and Magistretti (2006),

[…] this signifier would correspond to a modification of the synaptic efficacy (of some specific synapses) in relation to a unique, lived experience, which would be the signified. In this way, we could put on the same plane the modification of synaptic efficacy (the synaptic trace for neurobiology), the sign of perception (the psychic trace for Freud) and the signifier (for Lacan). These three terms (sign of perception, synaptic trace and signifier) would correspond to a meaning that is nothing more than the perception of the experience of external reality.

(p. 87)

They affirm that for Lacan the signifier and the synaptic imprint correspond, there are no deceptions. They draw a "bridge between the psychic trace and the synaptic trace established in the neuronal network" (p. 13), with which what the organism perceives is inscribed, signified, and remains available in memory: "the psychic [...] leaves material, concrete traces, consistent with experience" (p. 26). In such a way, the cure would be by biological decoding:

> In fact, it can be affirmed that the perception of external reality forms a sensory physiology, while the perception of internal reality constitutes a physiology of the unconscious [...] The task of psychoanalysis is to decode this physiology of the unconscious [...]
>
> (p. 165)

Neuronal plasticity would allow the subject to intervene via mnesics on his unconscious, connect with his memory, and become a unit with it, then the subject is his memory. This is how Dehaene exclaims it (cit. Simonet, 2019): "But we are our brains, of course!" (p. 8). So, Ansermet and Magistretti affirm that "plasticity is not only a concept, but a biological reality from which the notion of uniqueness of the subject arises" (2006, pp. 31–32). "In any case, the phenomenon of plasticity requires the psychoanalytic subject to think in the field of neuroscience itself" (p. 21).

> [...] we have tried to define a broad model that [...] is useful for understanding the biology of the unconscious [...] Plasticity would be neither more nor less than the mechanism by which each subject is unique and each brain is unique. Hence the title of this book: *To each his own brain!*
>
> (pp. 14–15)

For her part, Ariane Bazan (2011), a leading researcher in neuropsychoanalysis, doctor in biology, doctor in psychology, and professor of clinical psychology at the University of Brussels, Belgium, indicates that when someone speaks there would be no interval between words, the word would be directly the subject: "In spoken language there are no pauses between words" (p. 167). That is to say, for her, when someone speaks, memory speaks, therefore with a memory effort the unconscious would be reached, and with the signifier one would reach the signified. This assumes that the unconscious is in consciousness, you just have to identify it. Indeed, Kandel states: "One of the most surprising insights to emerge from the modern study of states of consciousness is that Sigmund Freud was right: we cannot understand consciousness without understanding that complex unconscious mental processes permeate conscious thought" (2018, p. 362). Let's translate this here: consciousness trains itself not to be excluded by memory. This is the point of view of Ellis and Beck (Ellis, 2000; Beck et al., 1983), central figures in the cognitive model for whom the therapist's task is to train while the patient is left to understand and collaborate. Hence the methods of self-knowledge or self-help, are sustained in this idea of correspondence. For example, the patient is asked to make a list of things he wants to do, likes and dislikes, or reasons for living.

Precisely, Davidovich and Winograd (2010) indicate that memory and unconscious integration is possible because brain mechanisms and unique processes are intrinsically interconnected. The memory retrieves everything, without scansions. Now, what distinguishes this idea from the preconscious? Kandel says that there would be no differences: "[...] the temporal lobe stores a particular type of unconscious information called the preconscious unconscious" (2007, p. 154). It is true that the brain is the site of the mind, the field of the secondary process that, according to Freud, governs the preconscious system, but not the unconscious.

The preconscious-unconscious equalization allows government and control efforts based on evidence-based scientific foundations. For example, at an airport in the United States, an investigation was carried out to identify – according to the manner of speaking, the shifting gaze and the tone of voice – potential terrorists passing through customs. The researchers carried out a series of neuroimaging studies and observed that there were certain areas of the brain that were activated, which consume a greater flow of oxygen and glucose in the blood when this type of behavior is carried out. Since the prefrontal cortex had been more activated at certain times and at other times it disappeared in the manifest, but the nucleus accumbens, areas of the midbrain and diencephalon were activated, it could be concluded that the conscious is linked to the prefrontal and the preconscious unconscious to the nucleus accumbens (Stagnaro, 2009).

This is the so-called evidence-based approach that erases the opportunity to create one's own reality for each subject. As Castanet (2023) says, psychoanalysis has a single means, the word of the analysand. The neurosciences want to show us the facts and in that they are similar to hysteria insofar as it is common for them to make themselves seen, to show the evidence. That's why Lacan says that "scientific discourse and hysterical discourse have almost the same structure" (2012f, p. 549). On the contrary, for psychoanalysis the facts are not a matter of reality, but are produced in the discourse of the speaker. "There are no other facts than those that the *parlêtre* recognizes as such by saying them" (Lacan, 2006a, p. 64).

Another neuroscientist is Adrian Johnston (2013). He has the thesis that nature does not determine (naturalistic indeterminism) and that this constitutes freedom. He justifies it by saying that the human cub, due to its defenseless state, needs the Other, and there the drive is ready to be educated.

Following Freud and Lacan, they point out that the prolonged period of premature helplessness in humans dooms them to the dominance of nurture over nature [...] the still-a-priori biological state of development of human infants lends support to the theme [...] of humans preprogrammed to be reprogrammed.

(Johnston, 2013, p. 66)

That is, according to Johnston, the premature condition of the human brain means that human nature is destined to denaturation via neuroplasticity, with which the subject becomes free and autonomous. Thus, neuroplasticity determines everything. So neuronal plasticity poses a naturalistic determinism, not a naturalistic indeterminism.

It can be noted that the concept of neuronal plasticity has moved current neurosciences. If we do a little history, according to Ibáñez (2017), neuronal plasticity was known about in 1975. The new thing that these neurosciences introduce is their application to the unconscious to change its causality. Indeed, when a plastic material is modified it no longer returns to its previous state. The reference would be surmountable in terms of training, of learning. The subject could move away from the origin due to the plastic quality of the neurons: "beyond the innate and any starting data, what is acquired through experience leaves a mark that transforms the previous" (Ansermet and Magistretti, 2006, p. 20). Even modifying the contingent: "Plasticity shows that the neural network remains open to change and contingency, adjustable by the event and the potentialities of the experience, which can always modify the previous state" (p. 20). That is: the subject could actively participate in the modification of his origin and his future. But this idea of modification from experience goes against Freud, who discovered something that does not change precisely in experience, something that is repeated, invariant, inert, once and for all, which Lacan called jouissance. Jouissance is what does not change, what is immutable, and what is stagnant, because it runs along a different path from that of the symbolic (Miller, 2011a). It can be argued whether jouissance is transformed at the end of an analysis. But from the outset jouissance is inertia. "Inertia is a word used by Lacan, it is a descriptive term to designate what does not change in experience" (Miller, 2006, p. 124). On the contrary, neuroscience wants to teach us that our brain is in permanent self-modification, that we are never the same brain, that we never use the same brain twice, that it is always changing, so how do they explain the repetition, recurrences, and repetitions of the same thoughts and actions?

Suppose that it is legitimized that the unconscious is finally in a neurological region, then synaptic stimulation would have to produce changes in the psychic, something that, as Yellati (2018) points out, was refuted by the famous Phineas Gage case in which brain stimulation (lesion) did not produce changes in the functions involved with the stimulated area. If it does not always happen that a function is modified by stimulation in the part of the brain with which it is related, what is the value of this neurobiological theory?

Procedural reality

Various studies have determined that the unconscious is procedural reality; it is better understood if we say: the unconscious is executive memory. Based on experiments with non-speaking beings, Kandel created the concept of "unconscious memory" stating that it is located in the amygdala: "The amygdala nucleus seems to participate in mediating both the unconscious emotional state and the conscious feeling" (2001a, p. 992). The amygdala is part of the limbic system. Its main function is the storage of emotional reactions essential for the survival of the individual. There resides a type of memory capable of seeking the balance of the living being, a capacity that would be related to the unconscious: "The amygdala – at the interface between sensory stimuli and the neurovegetative or endocrine systems that control homeostasis – seems to contribute strongly to the constitution of this unconscious internal reality" (Ansermet and Magistretti, 2006, pp. 193–194). Brenda Milner "discovered that in addition to conscious memory, in which the hippocampus intervenes, there is another, unconscious memory, whose seat is outside the hippocampus and the medial zone of the temporal lobe" (cit. Kandel, 2007, p. 158), thereby justifying the conception of this internal, unconscious, and procedural reality.

The links between procedural memory and the unconscious have been extensively studied in neuroscience. Vesa Talvitie, a doctor in psychology at the University of Helsinki who has worked on the Freudian concept of the unconscious, argues: "Freud held that patients suffered from reminiscences, and Erdelyi expressed it in terms of cognitive science: past events appear in procedural forms, but not in declarative memory" (2009, p. 30). In the same direction Delgado, Strawn, and Pedapati (2015) conceive the unconscious as implicit memory, that is, non-declarative: "Non-declarative memory refers to a non-conflicting dynamic unconscious" (p. 2). Solms (2017a) also supports the theory of the unconscious as the automation of non-declarative motor memories. Solms (2017b) locates the unconscious as what it is without representation, in automated action plans and basal ganglia memories. It is the automated unconscious and memory that does not require cognition. Then, Dall'Aglio (2019) qualifies this argument by extending the unconscious through the Lacanian real to his insistence on the level of affective instincts, which register activity prior to the development of the hippocampus and therefore function without representation. In this way, they say that the real is affectively conscious, without being cognitively represented in a declarative way. And Kandel (2018) places this association between the unconscious and implicit or procedural memory: "What makes implicit memory so mysterious, and the reason why we rarely pay attention to it, is that it is largely unconscious" (p. 176). In other words, "[…] the non-conscious and procedural memory are too often considered as equivalent to the unconscious" (2009b, p. 44). So, all these authors affirm that the actions that go through the procedural memory, mechanisms learned by the living being that do not need to be reviewed or reasoned to be carried out, are called unconscious. While other authors, for example, Eichenbaum, Cahill, Gluck, Hasselmo, Keil, and Williams (1999), affirm that the unconscious is not procedural memory, but declarative.

> […] the unconscious would be closer to the processes of declarative memory than to those of procedural memory; with the difference that, contrary to the memories of events and objects of external reality that are directly accessible to consciousness by mere memory, unconscious declarative memory implies the passage through the association processes facilitated by psychoanalytic work.
>
> (p. 44)

In short, the discussion is whether the unconscious is a procedural memory or a declarative memory, if it is the set of brain processes that support the motor mechanisms tending to restore balance or if it is the memory declared and identified (as in self-knowledge methods). In any case, memory allows us to study the brain as the foundation of what we are.

Internal and autonomous reality

The experience taken to the neural plane is founded on a perspective that sees through the glasses of interiority: seeing the interior of the body. For Ansermet and Magistretti (2006) the unconscious does not have a relationship with the outside world, but with the inside because it ignores what comes from the environment. "No one has a brain the same as another, there is not the same state of a brain at one moment and another in the life of a subject" (Bassols, 2011a, p. 89). Every day the brain is modifying itself and self-modification supposes dispensing with the Other, precisely because by obeying laws that come from within it is done autonomously (Bassols, 2012b; Castanet, 2023).

This notion contradicts the practice of psychoanalysis. Because if the Other is terminated, there is no longer a way to make someone take responsibility for their part in what they complain about. If we do without the Other, the human being ceases to be a social being, so animals that do not speak and beings that speak are the same. It is difficult to understand someone without the introduction of otherness (Bassols, 2013a).

Ansermet and Magistretti affirm that "the unconscious […] is constituted through the mechanisms of the plasticity of each subject, beyond the external reality actually lived" (2006, p. 96). "Thus, an action can be activated directly by unconscious internal reality without the intervention of an external stimulus; it can be generated from an unconscious phantasmic scenario in resonance with a somatic state" (p. 131). The somatic state, for example, hunger, stimulates unconscious processes.

> Our husband is divided between what comes from the current situation and what comes from an unconscious reality activated by chance by the pressure of a simple somatic state linked to hunger. The link between the two comes from the coincidence between the hunger he feels and the mother's historical relationship with food.
>
> (p. 153)

The unconscious is limited to survival mechanisms, governed by the principle of keeping the tension at the minimum possible level, caused by the internal organism toward the search for balance, an idea that is not very far from the adaptation and synthesis functions that we know from Freud for the Self. Says Langaney (2006): "The individual project, almost always unconscious, of every living being consists of keeping its structure in good condition for as long as possible, or if you like, maintaining its homeostasis" (p. 80).

In this sense, Gerard Edelman, an American biologist and Nobel Prize winner in physiology, and Giulio Tononi, an Italian psychiatrist, have demonstrated the "[...] way in which cortical and subcortical regions cooperate in order to connect conscious planning with unconscious routines" (2002, p. 39). "The impressive capacity of unconscious automatic processes in terms of speed and precision should not mask other characteristics that make conscious control vitally important" (p. 35). They link the executive areas with the brain nuclei of "unconscious cognition" (p. 108) charged with "physiological significance" (p. 108).

> First, the basal ganglia, the cerebellum, and the motor cortex are involved in the process of walking and in unconscious habitual procedures, such as turning on the tap to fill the glass of water. As we move, global maps send signals to our bodies, arms and legs, of which we are essentially unaware.
>
> (p. 123)

The autonomic mechanisms of the nervous system "provide a neurophysiological framework that allows us to understand how unconscious processes can affect the dynamic core and therefore influence learned and automatic routines" (p. 108). Even "[...] many aspects of speech are carried out through unconscious routines" (p. 120).

> We refer, for example, to the role of unconscious contexts, such as unconscious expectations and intentions, in the formation of conscious experience; to the conscious and unconscious regulation of attention; to the substrates and mechanisms of the Freudian unconscious.
>
> (p. 108)

Thus, for these authors, the unconscious and regulation go hand in hand. António Damásio himself, a Portuguese neuroscientist and doctor, subscribes to this notion under which body and mind are unified in the search for homeostasis. "Body and brain are inseparably integrated by biochemical and neural circuits that point to each other" (1994, p. 107). He wrote a book called *Descartes' Error* where he affirms that Descartes was wrong in thinking that there are *res extensa* and *res cogitans*. Damásio discards this dualism: he thinks that we do not have the brain, on the one hand, and the extensive body, on the other, but that we are a unity, not divided, undivided (individuals), which ignores the notion of the subject of the unconscious. In this sense, Kandel (2018) points out that the unconscious is linked to adaptation processes.

> The adaptive unconscious interprets information quickly, without us realizing it, making it vital to our survival. As we consciously focus on what's going on around us, the adaptive unconscious allows a part of our mind to keep track of what's going on elsewhere, to make sure we don't miss something important. The adaptive unconscious fulfills a number of functions, one of which is decision making.
>
> (p. 383)

The adaptive unconscious makes decisions to avoid harm. It is about the study of the "brain separated from the life of the individual" (Miller, 2015a, p. 182). When life is reduced to a set of mechanisms that save energy, it is enough for consciousness to select behaviors that can be automated and become unconscious so that the individual can rest (Cuñat, 2019). Example:

> A hysterically blind woman will flatly deny that she can see, but more often than not she will be able to avoid obstacles as if she were seeing them. Therefore, it is as if his ability to see has become unconscious or perhaps dissociated from his conscious self [...] Freud analyzed many similar cases [...] and suggested a link between these disorders and everyday phenomena such as [...] selective amnesia.
>
> (Solms and Solms, 2005, p. 42)

In a virtual seminar in the context of the IPA, Solms (2020) affirms that the non-declarative unconscious implies that patients suffer from feelings that would have to be readapted. Thus, the unconscious is the series of automatisms tending to homeostasis. On the contrary, for Freud the unconscious "is precisely what makes an unpredictable obstacle to all automatism, erasing the traces of the successive experiences of satisfaction – jouissance, Lacan would say – in which that subject hoped to represent himself" (Bassols, 2011a, pp. 133–134).

The concept of the unconscious being adapted to certain routines allows it to be operated via classical conditioning. Kandel affirms it: "[...] classical conditioning is an excellent example of how knowledge can go from being unconscious to being conscious" (2009b, p. 79). As Miller (2015a) says, the power of this perspective must be extracted from the need "for a gateway for the brain in culture, in cultural learning" (Miller, 2015a, p. 185). This is how the new biology of the mind, says Kandel (2009a), has allowed the unconscious to be understood as an element in a set of mechanisms of the nervous system:

> The new biology of mind that emerged at the end of the 20th century is based on the assumption that all of our mental processes are mediated by the brain, from the unconscious processes that guide our movements when we hit a golf ball to the complex creative processes that underlie it.
>
> (p. 42)

In the same school, Edelman and Tononi state that "active, but functionally isolated thalamocortical circuits may underlie certain aspects of the psychological unconscious – aspects that, as Sigmund Freud pointed out, share many of the hallmarks that define 'mental' states, except that they do not reach consciousness" (2002, p. 116). The thalamocortical nuclei are concerned with mediating general reactions of alerting, control, integrity care, and survival functions. Thus, the authors reduce the term unconscious to mere execution and thoughtless habituation, which clashes with one of the guiding principles of the analytic act: that the psychoanalyst authorizes one to distance oneself precisely from the habits to which the psychoanalysand submits outside of the session (Laurent, 2006).

The mental unconscious

The mentalization of the unconscious or the understanding of the concept of the unconscious in mental terms is one of the goals of the neurosciences. Various studies (Sulloway, 1992; Bazan, 2006, 2011; Bazan et al., 2007; Brakel and Shevrin, 2005; Solms, 2015, 2017b; Bazan et la., 2011; Solms and Turnbull, 2011; Shevrin, 2003; Redmond, 2015) have attempted to integrate psychoanalytic theory with neuroscience since the 1990s. They argue that Freud's initial attempts to ground psychoanalysis in biology were abandoned due to technical and conceptual limitations in science at the time. That is why they seek to verify Freud's ideas by integrating the psychoanalytic theory of the mind with the neuroscientific understanding of the brain. Thus the term "neuropsychoanalysis" is, for these authors, the culmination of past attempts to fulfill Freud's original attempts to ground psychoanalysis in a biologically based model of the mind.

In this direction, Kandel (2009b) argues that psychoanalysis is based on a perspective of the mind and therefore should add "the enormous baggage of knowledge about the biology of the brain and the control of behavior that has appeared in the last 50 years" (p. 71). Its objective is "to bring psychoanalysis closer to biology in general and to cognitive neuroscience in particular" (p. 68). For his part, Dall'Aglio (2019) argues that the brain can provide common ground for debating different constructs of psychoanalytic theory. Also, Solms and Solms (2005) return to Freud to affirm that "[...] psychoanalysis is both a clinical and a psychological discipline, which is dedicated precisely to the problem of correlating the phenomena of mental life with the structures and functions of the brain" (p. 6). Solms and Turnbull want to lay the foundations of the theory that locates the brain as a "useful set of anchor points from which to re-evaluate psychoanalytic concepts" (2002, pp. 42–43) and sign that psychoanalysis has to find in the brain the terms that identify the mental:

> [...] psychoanalytic observations about how the mind is altered by damage to different parts of the brain have enabled us to begin to build a coherent model of how the mental apparatus, as we understand it in psychoanalysis, is realized in anatomy and physiology, providing what we might call a new "physical" viewpoint in psychoanalytic metapsychology.
>
> (2011, p. 140)

Mark Solms, as indicated by Bazan (cit. Bazan and Zehetner, 2018), is the founder of the International Society of Neuropsychoanalysis. Neuropsychoanalysis is a current that emerged in the United States in the year 1990. At that time, a group of neuroscientists began to question psychoanalysis and came together to ensure the consistency of its scientificity (Lardjane, 2019). Ten years later, Solms and Solms (2005) developed this direction, taking Freud's work to legitimize that they had made "particularly notable progress in this regard in relation to the psychoanalytic theory of dreams, through the use of multiple converging methods" (p. 140) and state that "it has been gratifying to rediscover the Freudian conception of dreams in the neurodynamics of the sleeping brain" (p. 140).

Mark Solms has earned a reputation for his demonstration that the neuroscientific basis of dreams fits very well with Freud's general model of dream fulfillment (Bazan and Zehetner, 2018). His central thesis is that the dynamic localization of the unconscious within the clinical-anatomical method with which dynamic changes in the mental apparatus can be correlated with the locations of the brain (Solms and Solms, 2005; Dall'Aglio, 2020a). Solms's contribution makes it possible to defend neuropsychoanalysis from the criticism it receives regarding the fact that it supported its method based on brain images. It allows us to show a type of neuropsychoanalytic investigation that does not submit to the all-encompassing promise of brain imaging by attending to the unconscious as revealed in dynamic changes through speech. In this way, it seeks to uproot itself from the classical position of neuroscience, from its neurobiological vestiges, to uproot itself from a nosological claim based on locating the dimension in a certain region of the body. The work of Pina (2008), a renowned Catalan psychiatrist, also follows this path, combining biology with psychoanalysis to go beyond the modern dimensional orientation of psychiatry and collaborating in the necessary change that makes it possible to renew a single categorical orientation in psychopathological work.

The idea that Solms (2015, 2017a, 2017b) postulates is that the mind can be admitted as being located in the brain. He argues that the unconscious is the inaccessible part of the mind, it is equated to the non-declarative memories of the upper limb and that is why the unconscious is located in the basal ganglia. Memories that are inaccessible to consciousness, that cannot be represented in working memory, are mental. This mental unconscious would be located, says Solms, in "the lower brain processes" (2015, p. 144), against the grain of the higher processes that would be the conscious mind. Thus, Solms and Solms (2005) state that it is possible "to combine psychoanalysis with neuroscience, on a solid clinical basis [...]" (p. 89).

For his part, Redmond (2015) argues that psychoanalytic theory is associated with the conceptualization of "the mind" whose study is based on the use of brain science methodologies such as imaging technologies and techniques to measure implicit cognitive processes. For this author, implicit cognitive processes constitute the object of study of psychoanalysis, which would be dedicated to studying the dynamic nature of the unconscious mentality and its underlying neuronal organization. Of course, such an objective can have no other purpose than to challenge Freud's hypotheses and theories, and to develop new theories driven by psychoanalytic ideas about mental functions.

Dall'Aglio (2021) argues that psychoanalysis deals with the mind. He proposes a neuropsychoanalytic model where the unconscious is reconsidered as a dynamic effect of separation between mental levels: "psychoanalysis operates at the level of mental meanings" (p. 126), that is, the "psychoanalytic study of subjectivity provides unique information about the mind" (p. 128).

Kandel claims that "Freud had documented the importance of unconscious thought" (2018, p. 236) and Talvitie (2009) argues that "neuropsychoanalysis focuses on Freudian thought" (p. 96). Let's translate: the Freudian everything, everything Freud worked on, all his thought, "his" thought, has a center, all Freud is reduced to a center: neuronal data.

Even Talvitie is categorical: "Many researchers, Freud being the best known of them, have developed psychological explanations: dreams tell us something about the unconscious mind" (p. 20). If the mind is the psychic substance and this is not only the conscious, then a part of the mind is unconscious. Talvitie supports this syllogism with this phrase from Freud: "There is a right to respond that the conventional equating of the psychic with the conscious is entirely inadequate" (2012b, p. 164). Talvitie assures that "the mental unconscious has been, and continues to be, the cornerstone of Freud's legacy" (2009, p. 32). Likewise, Solms argues: "Freud says, very clearly in many places, that the real nature of the mind is unconscious. He uses the phrase 'the mind itself'" (2015, p. 19). With this, "the unconscious is identification" (p. 8), and there are "unconscious cognitions" (p. 29) that allow it to be represented.

Dunja Voos, a German psychotherapist, affirms that "Sigmund Freud explored the unconscious mental conflicts that his patients presented in order to free them from fear and anguish" (2013, p. 23). All these authors support the mentalization of the Freudian unconscious. So much so that Solms and Gamwell (2006) argue that the change in Freudian doctrine is not from neuroscience to psychoanalysis, but from an unconscious as a system of neurons to an unconscious as a mental system. They justify this with the drawings that Freud makes, taking as reference the first diagram of the psychic apparatus in "Letter 52" and the first *The Interpretation of Dreams*:

This was the breakthrough in psychoanalysis proper. A comparison between Freud's later neurological drawings and his first metapsychological drawing reveals unequivocally that little had really changed. The drawings were almost identical; the neuronal systems were simply renamed mental systems. The drawings still represented the same thing, namely, the sequence of changes that occur during the processing of stimuli, as they proceed from the perceptual to the motor end of the apparatus.

(pp. 16–17)

Edelman and Tononi (2002) also produced a work in which they titled a chapter "Unconscious Perception". It does not take much development to capture what is justified there, the unconscious is data perceived and processed in a mental system by the central nervous system:

In the laboratory, subliminal perception – now commonly referred to as unconscious perception – is often demonstrated by the presentation of stimuli that are too weak, short, or loud to be consciously perceived, but are sufficient to prime or bias the subject's ability to perform lexical decision tasks or similar tests.

(p. 44)

These "unconscious neural processes" (p. 107) are subliminal, that is, the authors give the individual a mental intention to carry out an action that contains a message below (sub) the normal (liminal) limits of perception. For this reason, Talvitie (2009) understands that when Freud said that the unconscious makes sense, it means that the unconscious is mental insofar as an underlying intention can be known:

> In the scope of this work, all this means that the statement "The unconscious is essentially mental" is (almost) identical to the statement "The unconscious possesses original intentionality". Therefore, if the latter holds, the unconscious is essentially mental.
>
> (p. 83)

There would be an intention already realized that can be mentalized. How to attribute to each analysand what returns to him from what he enunciates if the truth is identical to what is said? "What Freud referred to is the fact that the mind can be the first and the last subject" (Solms, 2015, p. 3). It would then remain to make the subject's intentionality conscious. Pinker (2001) says that this "meaning of the term consciousness no doubt also encompasses Freud's distinction between the conscious and the unconscious mind" (p. 181). Hence, for these neurosciences, psychoanalysis is a model anchored in the unearthing of the unconscious.

> We define the traditional psychology of a person as that which is based on traditional psychoanalytic concepts and the form of technique that emphasizes the role of the psychotherapist as an objective observer (a persona) of the patient's ego defenses (the symptoms) and the discoverer of the truth about the patient's intrapsychic conflicts and object relations (the patient's inner life).
>
> (Delgado, Strawn, and Pedapati, 2015, p. 1)

Note that Delgado, Strawn, and Pedapati (2015) mention the "traditional psychology of a single person". If tradition is what is repeated, then what did Freud clarify? The "one-person" part is because they also theorize a method with Freudian foundations of more than one person, and more than one person in an analytical device is already a group. Freud had initiated the invitation to freedom of speech, but with one caveat: it excludes that there are more than two people. Psychoanalysis is not a group practice. As Miller (1996c) says, when there is a group there is identification with thought, harmony, convergence, less tension, and duration are achieved. The group lives under the exclusive regime of the same. "I am like you", "the same thing happens to me as you". Group phenomena precipitate in the illusion that thousands of mouths say and must say the same thing. The most terrible disease of a group is that it is a prisoner of its own unknown jouissance. It is to quote the word of the Other erasing the enunciation of the subject. Freud, in 1908, formed the psychoanalytic society to separate the group of psychoanalysts from a scientific society founded on a common practice, since it is not with the crowds that one fights against the malaise of the master's identifications.

> If I had gone into psychoanalysis, I would have spent a good part of my life listening to patients talk about themselves: about their dreams and their daytime memories, their conflicts and their desires [...] In the Grundfest laboratory, I soon realized that in order to understand the functioning of the brain, I had to learn to listen to the neurons, to interpret the electrical signals that are the foundation of all mental life.
>
> (Kandel, 2007, p. 97)

For Kandel, neurons speak. That's why psychoanalysis didn't work for him. In his latest book he insists that for Freud the unconscious is comparable to the mental: "[...] as previous scientists discovered and as Sigmund Freud had emphasized, our conscious perceptions, thoughts and actions are informed by unconscious mental processes" (2018, p. 16). Thus, the notion of the mental unconscious reduces psychotherapeutic work to the training of mechanisms of the self in order to achieve a mature vision of it:

> In a traditional one-person model of psychology, insight is the ultimate goal of treatment: it aims to help the patient develop insight into his repressed and unconscious conflicts and learn to use mature ego defense mechanisms.
>
> (Delgado, Strawn, and Pedapati, 2015, p. 148)

The term "maturity" comes from biology. It supposes an evolutionary cycle where the being has to go through certain stages, the same for all, and there is only one path, that of progress, where the object is predetermined. At each stage, it is known what the being has to achieve in order to grow in its species. Analogous to the idea of pedagogy, of degrees toward obtaining a degree, training under the aegis of the norm absolutely erases the notion of symptom and singularity. There is no possibility of someone getting involved in this evolution because mature development is destiny. It is an erasure of the subject, which has the effect, affirms Miller (2004c), of the resurgence, on a large scale, of the passage to the act. If the subject is ignored, the passage to the act is imminent. There emerges with emphasis the adaptation of being. Adaptation is the model. Kandel (2018) suggests that the last version that Freud established is an adaptive unconscious: "Freud completed his structural model of the mind by adding a third component, the preconscious unconscious, which is now called the adaptive unconscious" (p. 364).

> [...] when we speak, we know the essence of what we are going to say, although we do not know precisely what we are going to say until we say it. Similarly, when we look at a face, we don't consciously see two eyes, two eyebrows, two ears, and a mouth and say, "Oh yeah, that's so-and-so". Recognition just comes

to us. Such high-level adaptive thinking takes place in Freud's preconscious unconscious.

(p. 372)

If the individual when he speaks knows the essence of what he says, why does he go to the analyst? As Laurent (2005) points out, the unconscious, for these authors, is a process that is not yet mental, but if the unconscious were finally mental, where would experience be?

Between two neurobiological theories

Initially, the study of the brain was based on the genetic theory that functions were inherited. However, the genetic theory precludes thinking about rehabilitation. From there arose neurology as a branch of medicine and the clinical anatomy method that allows the study of brain lesions to find out their symptoms and their functions. The assumption that the brain can present pathologies associated with its tissues and that damage to parts of the brain produces different clinical syndromes that have immediate effects on the patient's mind was the basis of clinical neurology. Their goal was to describe which lesions should be correlated with which symptoms.

Thus, since 1860, research on the relationship between psychological processes and the brain organization that makes them possible began to be validated (Mías, 2008). The first milestone was in 1861 when the French neurologist Paul Broca demonstrated that damage to a specific part of the brain caused loss of the ability to speak despite normal functional ability in the physical organs of articulation, a syndrome known as Broca's aphasia. This showed the locality, the physical seat of a symbolic function like speaking and thus the localization of language in the brain became evident. Years later, in 1874, Carl Wernicke, a German neurologist, showed that damage to another specific part of the brain caused loss of the ability to understand language, a syndrome known as Wernicke's aphasia. This was another achievement for the nascent neurosciences in localizing language to another specific site in the brain (Solms and Solms, 2005; Jones, 1981).

Such discoveries were followed by a long series of investigations regarding anatomical-clinical correlations in relation to a variety of other functions such as writing, visual recognition, attention, memory, language, among others. For example, for Bazan (2011) "there is, in human beings, an organized brain circuit that deals specifically with words, with language as such" (p. 166). While for Lacan "language is not located in the nervous system [...]" (Bassols, 2011a, p. 128). Language precedes the subject in its existence so it cannot be in an organ of the body. Language is not in the materiality in which the subject incarnates, in physical reality, it is not there, it is outside, it is in the field of the Other. "Language can in no way be a neural organ [...] The problem of language is inseparable from this place of the Other that precedes the subject in his birth" (p. 92). This place of the Other is where Freud began by situating the discourse of the unconscious (Lacan, 2012b). Language is what travels in words, it does not have a fixed place, it does not have a material support and it is what will activate the subjective functions in the biological support in question (Bassols, 2011b).

> In other words, and if you allow me to exceed, my central nervous system right now has its extension in this wiring that goes from the microphones to the speakers that are just as supportive of language as my body so that my word reaches you. There is no need to resort to any organicist theory of language to find its place.
>
> (Bassols, 2011a, p. 92)

However, research on injured patients fascinated the neurosciences. They have been able to identify the damaged neural area that would be necessary for the underlying function. As Foucault says (cit. Castanet, 2023) this method of anatomical-clinical correlation constitutes the basis of a medicine that is presented as positive, where "medical observation then leads to the disappearance of the being of the disease" (p. 23).

However, the association between mental functions and brain lesions began to contain a difficulty that the linguist Roman Jakobson (cit. Bassols, 2011a) highlighted when he verified that the aphasias produced by cerebral lesions "[...] followed the symbolic laws of the language of metaphor and metonymy [...]" (p. 94). In other words, aphasias are not due exclusively to the organic region involved, but to laws of the signifier.

Neurobiology then responded with neuropsychoanalysis, which located the cause of functions no longer in one place, but in the framework, in the organization of brain structures, that is, in activity. The function is no longer located punctually in a specific structure, but in its complex organization, with which genetics are modified by pharmacological and psychotherapeutic treatments that would act at the same level of synaptic transformation made possible by neuronal plasticity (Kandel, 2009b). This is the current theoretical state of neurosciences.

Recent technologies allow the brain to be opened in the living state. There human functions are studied in real time because active regions can be seen in certain types of mental processes, in specific circumstances, and then see differences in people with different diseases, observing how their brains, damaged, are working. Direct informative results have been obtained for having opened the inscrutable black box that was the brain for science (Barondes, 2004). "Based on this, a wide range of psychological faculties were located in a mosaic of so-called 'centers' on the surface of the brain" (Solms and Solms, 2005, p. 10). Among such faculties is, according to Solms (2015), the unconscious that "is located on the surface of the brain" (p. 89).

In this sense, Solms and Solms (2005) have the thesis that Freud's contribution was to establish that hysterical paralyzes are considered psychological disturbances of a functional cerebral complexity. They take paragraphs from *The Interpretation of Dreams* to state that Freud is definitely in favor of functional theory, rejecting classical locationism and its anatomical factor as the cause of the disturbances.

> We want to put aside completely that the psychic apparatus in question here is also known to us as an anatomical preparation, and we will take the greatest care not to fall into the temptation of determining this psychic locality as if it were anatomical.
>
> (Freud, 2012c, p. 529)

And later they present the following paragraph:

> We will avoid any abuse of this mode of figuration if we remember that representations, thoughts and, in general, psychic products cannot be located within organic elements of the nervous system, but, so to speak, between them [...]
>
> (p. 599)

They also quote from "Fragment of analysis of a case of hysteria":

> [...] the theory in no way fails to point to the organic bases of neurosis, although it does not look for them in an anatomo-pathological alteration; it is possible to expect to find a chemical alteration, but, since it is not yet apprehensible, the theory provisionally substitutes it for the organic function.
>
> (2011ll, p. 99)

And in the "18th conference" they take: "Our psychic topic provisionally has nothing to do with anatomy; it refers to regions of the psychic apparatus, wherever they are located within the body, and not to anatomical locations" (2011l, p. 170). With these quotes, Solms and Solms argue that Freud's contribution was the move from anatomical theory to functional theory. These are isolated quotes from between 1900 and 1916, and therefore it does not represent all of Freud's work. They do not account for the continuity or discontinuity of this idea throughout his work. On the contrary, Mark Solms (2015; Solms and Turnbull, 2011) considers that the "Project of a psychology for neurologists" was Freud's most valuable contribution because it accounts for the transition from the anatomical to the functional, but he did not publish it because it had little scientific progress. If we follow these neuroscientific arguments then Freud's intention would have been to name the theory and method he discovered as "Neuropsychoanalysis", but he would never have said it that way. We therefore owe this new discovery to current psychoanalytic neurosciences.

If we read Freud, but not in parts, we see that he is reformulating himself all the time. The first Freud is not the same as the last Freud. He begins by saying that the dream is the fulfillment of a wish and ends by saying that in addition to that there is something beyond, that the dream repeats a traumatic experience; it begins by saying that sleep is a pleasure realized and ends by postulating the need for self-punishment. Such reformulations demonstrate, at least, that it is suspicious to postulate Freud's intention based on only some parts of his work.

From their perspective, Solms and Solms (2005) affirm that Freud never stopped considering the psychic processes associated with natural energies as he had learned in medical school with Helmholtz, and that he never moved from that perspective. "All he abandoned was the narrow localizationist notion that psychological processes, which have complex, dynamic, and functional organizations, can be concretely located in discrete anatomical structures" (p. 21). So, Solms argues that Freud breaks away from the anatomical localizationism that Charcot taught him to respect and adopts functional theory; that Freud made the path from histology to neurology, and from neurology to neuropsychoanalysis; and that there is a place in the brain to locate functional processes such as the unconscious (Solms, 2004, 2017a).

Several neuroscientists share this thesis. For example, Deneke (2006) affirms that "Sigmund Freud today would possibly be one of the first to use the knowledge of neurobiology to rework in depth his conceptions about the formation of the psychic apparatus" (p. 71). Talvitie (2009) maintains the idea that the project was Freud's greatest contribution when passing the psychic apparatus through neurophysiology, although there are certain oddities in this that could not be explained by Freud, but "current neuropsychoanalysis has revived interest in explaining these oddities neurophysiologically" (p. 23). Neurophysiology, neurofunctionality, and neurodynamics are some of the terms chosen by the aforementioned neuroscientists to interpret certain parts of Freud's work that apparently seem strange to them. Thus, according to these neurosciences, Freud's operation is reduced to overcoming anatomical localizationism and adopting functionalism to locate the unconscious. "In the absence of finding a place, we speak of processes, which in turn involve not only the brain, but also the central nervous system, all the peripheral systems" (Bassols, 2011a, p. 90). Now, if we assume that there is a consensus in admitting the functional location of the unconscious in the brain, Freud would not agree either!

> Here is a thesis whose negative side is well understood: it is equivalent to affirming that in the autopsy no appreciable tissue changes will be found [...] I am quite sure that many of those who read Charcot's works believe that a dynamic lesion is a lesion whose traces are no longer found in the corpse, such as edema, anemia, or active hyperemia. But, although they do not necessarily persist after death, even if they are slight and fleeting, these are genuine organic injuries.
>
> (2011j, pp. 205–206)

In other words, if there is a functional alteration, we will necessarily find anatomical lesions that affect larger regions. Dynamic lesions imply tissue imprints. The difference between anatomical and vascular, dynamic or functional theory is then doubtful. Therefore, functional theory is another localizationist, anatomical theory.

The difference that Freud (2011a, 2011j, 2011o) reveals is between the psychic causation and the biological causation of paralysis. He says that while hysterical paralyses are capable of the maximum intensity and the clearest isolation, organic paralyses do not unite intensity and isolation. For example, a monoplegia of the arm, of organic cause, is almost never absolutely specific, but can extend to a paresis of the face and legs. On the other hand, a hysterical paralysis can be absolute, individual, and only of the arm. In hysterical paralysis, each muscle, each muscle fiber can be paralyzed individually and in isolation, that is, they are not in complete paralysis. On the other hand, in organic paralysis it is always a condition that attacks a wide area of the periphery, a limb, a segment of it, a completely complicated motor apparatus, but it never affects a muscle individually. Even the central segment in a hysterical paralysis can be more affected than the peripheral segment, contrary to all paralysis rules in clinical neurology. The central area may be affected, but there is sensitivity in the periphery. They are paralyses that offer themselves isolated to the entire system, zones that are isolated, subtracted from the system and paralyzed independently. That is the particularity that distinguishes hysteria from a palsy with an organic cerebral cause.

> If in this way hysterical paralysis approaches cerebral palsy, and in particular cortical palsy, which presents a greater facility for dissociation, it is still distinguished from it by some important characteristics. In the first place, it is not subject to the rule, constant in organic cerebral palsy, that the peripheral segment is always more affected than the central segment. In hysteria, the back or thigh may be more paralyzed than the hand or foot. The movements can reach the fingers while the central segment is still absolutely inert. It is not the least difficult to artificially produce an isolated paralysis of the thigh, leg, etc., and very often it is possible to find these isolated paralysis in the clinic, in contradiction with the rules of organic cerebral palsy.
>
> (Freud, 2011j, p. 200)

Hysterical paralysis can become absolute and remain delimited at the same time. While organic paralysis can cause, for example, monoplegia, and not necessarily hemiplegia, it will still willingly affect surrounding regions, that is, it cannot be delimited.

> If the arm is paralyzed as a result of an organic cortical lesion, there is almost always a minor concomitant condition of the face and leg as well, and if this complication does not appear at a given time, it was nonetheless present at the onset of the condition.
>
> (p. 202)

In contrast, a hysterical paralysis may inhibit a particular area without necessarily affecting nearby areas or affecting the sensation of the whole half of the body.

> For example, it affects the arm in an exclusive way, but there are no traces on the leg or face. Furthermore, at the level of the arm it is as strong as a paralysis

can be, and that makes a marked difference from organic paralysis, a difference that is very sobering.

(p. 202)

What Freud is saying is not that psychological disturbances are caused by a functional cerebral complexity, but that he is highlighting the idea of hysterical symptoms and says that it is what can appear in isolation. Anesthesia and hysterical paralysis "are not accompanied by the general phenomena that in cases of organic lesions attest to the encephalic condition" (2011a, p. 53); "next to a paralyzed arm, a fully intact leg on the same side" (p. 53). That is, hysterical hemianesthesia, the insensibility of one side of the body, is not total. Hysteria admits the conjunction between a maximum development of the disturbance (excessive pain) "and its sharpest demarcation" (p. 53); for this reason, "hysterical symptoms are mobile in a way that refutes all conjecture in advance of material injury" (2011a, p. 53). This contradicts the conditions caused by organic lesions in brain regions.

I affirm, on the contrary, that the lesion of hysterical paralysis must be completely independent of the anatomy of the nervous system, since hysteria behaves in its paralysis and other manifestations as if anatomy did not exist, or as if it had no knowledge of it [...] Hysteria is ignorant of the distribution of nerves.

(Freud, 2011j, p. 206)

By not obeying the common rule of biology, hysteria behaves as if the nervous system did not exist. Freud gives a causality other than the organic one and decentralizes the localizationist intention of the neurosciences. It is true that with the functional theory neuropsychoanalysis takes a step further than what Broca and Wernicke did, eminent twenty-somethings in neuroscience who ascribed more to an anatomical theory, but, basically, they continue with the same reference in neural matter, therefore it is a discussion within biology, we could say that the discussion is neurophysiology versus neuroanatomy. Is this enough to move into the field of psychoanalysis? Solms may answer yes, because for him neurofunctional theory has managed to clarify Freud:

Recent neurological mapping fits Freud's description. The central region of the brain stem and the limbic system ¾ responsible for instincts and drives ¾ corresponds to Freud's id. The ventral frontal region that controls selective inhibition, the dorsal frontal region that controls conscious thoughts, and the posterior cortex that perceives the outside world correspond to the ego and superego.

(Solms, cit. Laurent, 2008, p. 8, 2005, p. 65)

In this sense, Dehaene (cit. Castanet, 2023) affirms that if we stimulate the insula with electrodes, which "is a fold of the cortex deeply buried under a part of the temporal and frontal lobes" (p. 84), we obtain effects of suffocation, heat, nausea, or vertigo; if we stimulate in the direction of the subthalamic nucleus we obtain a depressed state, crying, a monotonous voice, and a gloomy body posture; if we stimulate the parietal lobe we obtain levitation effects, of feeling as if floating in the air. And so much research could be done on injured brains. This means that the neuroscientific intention is always localizationist. Localization is an undoubted trend in the neurosciences. It makes it possible to ensure that all thought comes from brain activity; it frees the subject from having to know his part: there is nothing wrong with me, what is wrong is with my brain! The pathology can be eliminated and the individual moved to the happy mean point on the normal distribution curve.

The functional theory is based on the changes in blood circulation in these various structures of the brain observed when neuronal activity increases. The thing happens entirely in the notion of activity, from neuron to neuron, the foundation of the concept of neuronal plasticity, invented to get out of its confinement of being reduced to natural science laboratories and neurosurgery medical departments, but it is still a localizationist theory, it does not gather arguments to go to the field of psychoanalysis. And if it did, it would be a callous passage, strangled by biology.

It is still doubtful whether they have been surpassed. Fajnwaks (2019) affirms that neurosciences have produced a return of materialism because the tendency to correlate subjective phenomena with the physical goes back to the development of neurology in the 19th century.

> In the 19th century, attempts were already made to locate the organ of mental synthesis and to identify what Aristotle called common sense. Dehaene quotes Avicenna, who in the year one thousand located common sense not far from the frontal cortex, although without having our means of investigation.
>
> (Miller, 2015a, p. 183)

It seems that the effort to improve comes from the side of using Freud's work to solve a dilemma that has troubled the very field of biology for thousands of years. That is why they have changed their language, into another language, which allows university students, biologists, scientists, and the whole universe to have the illusion that they are talking about something in common (Bassols, Laurent, and Berenguer, 2006).

Drive, jouissance, and motor activation

Solms and Turnbull (2011) argue that the psychoanalytic community would have to go back to Freud to understand drive from conduction theory: "drives may play a more substantial role in mental life than previously thought" (p. 142). The term driving refers to mechanics or physics from which the impulse is a kind of force of movement, of action toward something. But this clashes, at some point, with Freud's conceptions of the drive. For Freud (2011d) the drive is a representative agency, and by saying "representative" he is saying that there is no relationship with the thing. He says that the drive is a representative agent in the psyche of a source of stimuli that comes from the intrasomatic, that is to say, stimuli that push from within, that emanate from inside the body, but he does not say that it is action in direct relation with the thing.

On the contrary, the driving theory of the drive maintains, as Bazan and Detandt (2013) point out, that, in the course of evolution, specific actions had to be superimposed on internal needs for living beings to function efficiently, for example, respiration for oxygen, digestion for food, hydration for water, reproduction for sexual objects, etc. From this perspective, a first physiological understanding of the Freudian concept of drive would be the dynamics by which a bodily tension, originating from a need in the internal body, mobilizes the external body and instigates it to the appropriate action that cancels the tension. There would be a certain end to cancel all needs and relieve tension. In this sense, the drive is what is appropriate to an end, invariant among all individuals of the same species, where the goal is fixed, welded to the end of the adaptation. On the other hand, for Freud (2011d, 2011g) the drive does not have a determined object, it is not welded to any object, it can be transformed, and it is variable. For this reason, Freud says that the drive goes beyond the physiological data. "Now we have obtained material to distinguish between drive stimuli and other (physiological) stimuli [...]" (Freud, 2012g, p. 114).

This distinction that Freud makes is fundamental. Instinct is associated with a stimulus. The stimulus operates in a single blow, for example, a whiplash, a blow to the body, feeling cold or hot, or a disturbing sound, which causes the living being to react to calm the tension. Instead, the drive, says Freud, is a constant force from within that cannot be exhausted. "The drive [...] does not act as a momentary shock force, but always as a constant force. Since it does not attack from the outside, but from the inside of the body, a flight cannot be worth anything against it" (2012g, p. 114). A flight, like the reflex act, can be useless to cancel its source. The drive always presses, it never ends, it has no end. If the subject is thirsty, it is canceled with water, that is an instinct. Now, if the subject is thirsty and wants whiskey, or Coca-Cola, it is no longer thirst, but a satisfaction. The drive is distinguished from instinct "by the fact that, in its Freudian definition, it does not subside, it does not cease" (Miller, 2021d, p. 212). It becomes impossible to reconcile that drive force and if neurobiology seeks cohesion, how could it be combined with psychoanalysis?

From the neuro perspective, the drive is associated with motor activation in search of a goal object that ends the tension coming from the stimulus. In this sense, they find the link with the so-called NAS-DA reward search system. This dopaminergic system is made up of the mesolimbic pathway, in which the ventral tegmental area (VTA) innervates the covering of the nucleus accumbens (NAS), which is part of the striatum (basal ganglia). When this system is stimulated, organisms display the most energetic exploration and search behaviors that an animal is capable of exhibiting, for example, stimulated rats move excitedly, sniffing vigorously, stopping sometimes to investigate various nooks and crannies in their environment (Panksepp, 1998; Bazan and Detandt, 2013).

In this way, motivations given by neural circuits are produced that Panksepp (1998) has classified into seven categories, seven elaborated affective instincts: search, fear, panic, game, care, lust, and rage. These instincts, according to Dall'Aglio (2021), "are common to all mammals and are present at birth" (p. 131), so they do not require language. An exhaustion of food will lead to the release of dopamine and this to motor mobilization. Without the synaptic energy of these circuits we would be still. Thus, the authors (Bazan and Detandt, 2013; Panksepp, 1998; Shevrin, 2003; Dall'Aglio, 2019) provide us with a complete natural description of subjectivity by stating that these dopaminergic circuits, which arise from the midbrain nuclei, are necessary for body survival. They have been able to show that damage to this system produces general behavioral quiescence, while abundant activation of dopamine synapses makes a person feel like they can do anything.

In this way, the Freudian concept of drive is related to bodily imbalances, properly due to internal needs, while they induce the activation of the search system, that is, they induce the release of dopamine. Therefore, the drive is located in the brain stem and in the diencephalon where "need detectors are contained" (Dall'Aglio, 2019, p. 28). And there is nothing left but to define it as the search for an external object that constitutes the goal that provides the consummatory satisfaction of the need. With which, jouissance and pleasure are not distinguished. In fact, Bazan and Detandt (2013) ask "how can we now situate the proposed difference between pleasure and jouissance in this physiological model?" (p. 6).

Ariane Bazan is a fellow at the Stellenbosch Institute for Advanced Study in South Africa and Sandrine Detandt is a researcher at the Fonds de la Recherche Scientifique in Belgium. They have written a majestic article in which they have proposed the neurophysiology of the Lacanian concept of jouissance. It is a difficult work where a parallelism between the concept of jouissance and the dopaminergic system is proposed. They define jouissance as "an accumulation of bodily tension, fuel for action, in a continuous balance between reward and anguish, and both marking the physiology of the body [...]" (p. 1).

For the authors, jouissance is a characteristic of behavior that has enormous adaptive importance because it allows organisms to make an effort to overcome obstacles. This concept of enjoyment "can be linked to actions that (once) were adequate not only to obtain pleasure (gratifying situations), but also to avoid displeasure (aversive situations)" (p. 8). The release of dopamine generally does not occur during the consummatory phase, but before it, inducing a state of motor tension that leads the organism to move toward the rewarding stimulus. This release of dopamine defines the "jouissance within the framework of the experience of satisfaction, that is, the bodily motor tension instigated by wishful activation" (p. 9). In this sense, various studies (Bazan and Detandt, 2013; Shevrin, 2003; Dall'Aglio, 2020b; Yue and Cole, 1992; Decety, 1996; Gallese, 2000) state that enjoyment is the benefit obtained from the motor tension underlying the action that was once adequate to relieve tension, that is, enjoyment is a state of motor activation. Any action intention, whether executed, thought, or imagined, leads to a slight increase in muscle tension. With this implication of the body, they match the concept of jouissance posed by Lacan: "jouissance is in the motor mobilization or use of the body, that is, in the motor mobilization of those pathways of action that were (once) adequate to provide pleasure" (Bazan and Detandt, 2013, p. 2). Given the need, the investment of that effective model of tension resolution is activated and "this motor tension would then be equivalent to the Lacanian concept of jouissance" (p. 3). Therefore, for this notion, what is repeated are situations that generate pleasure.

In fact, an article written by Bazan, Detandt, and Van de Vijver (2017) has appeared, in which they try to locate neurophysiological foundations to sustain that repetition establishes a field related to the homeostatic search for pleasure. They indicate that even having a beyond, Freud continues to uphold life tending toward balance and that Lacan also affirms homeostatic organic demands. According to this study, what Freud classified as a traumatic experience is the reception for the first time by the organism of some essential object for survival. For example, the first air that enters the respiratory system or the first milk that enters the digestive system, that is what is traumatic for Freud according to these authors. They explain contingency as the surprise encounter between the incoming milk and the organism, causing the organism to react with sucking movements that it has not learned. The entrance of the milk is inscribed as a mark of the event that has taken the individual by surprise. This brand, the authors affirm, is "witness of a subjective choice" (p. 5) because "the little human is surprised by the entrance of milk, he could well have chosen not to drink" (p. 5). The position of the unconscious is determined by the neurophysiological exchange between the organism and the environment, by the mechanisms of entry and exit, things go in here and go out there, which supposes a total translation of what goes in to what comes out. In this way, surprise would not be exclusive to humans: "this logical time of being surprised is not exclusive to human beings, but rather, as we will explain later, it has to be assumed in vertebrates in general as well" (p. 5).

Desire, according to Bazan and Detandt (2013), also depends on motor activation. They rely on Berridge and Robinson's experiments with rats, with which they have established a distinction between wanting and liking. The liking reactions will be measured based on the rat's facial reactions and the wanting will be measured based on the amount of motor activation that the organism is willing to invest to obtain the reward. Easy: if the animal does not move, it is because it does not want to. Desire and behavior are associated. Thus, if dopaminergic systems are involved in desire, stimulating dopamine neurotransmission will be sufficient to produce desire effects in humans. So that the desire and the drive "have some similarities insofar as both refer to the willingness to engage in a motor behavioral effort" (Bazan and Detandt, 2013, p. 7).

Thus, the dopaminergic systems embody the physiological architecture of Freud's concept of drive and Lacan's concept of jouissance, remaining as purely somatic concepts, which results in a triumph over castration. Lacan placed this as a rejection of the unconscious. Indeed, regardless of the subjective state, when the NAS-DA circuits are activated, the animal is put into a state of intense motor activity. According to Panksepp (1998), the human being is energetically active. This exclusively implies a motor factor: he works not by desire, but by the laws of this system, that is, by motor pressure. He directs behavior toward a goal capable of restoring homeostasis. It is a perspective focused on action rather than stimulus, with which they say they have overcome centuries of behaviorism. Action-focused because the dopaminergic system drives thirsty animals to water, cold to heat, hungry to food, and arouses the system to steer them toward opportunities for orgasmic gratification. Thus, the individual enters a kind of manic exaltation that supposes, as indicated by Fridman and Millas (2005a, 2005b), the death of the subject.

Manic exaltation is one of the ways in which subjective death occurs. This is the interval loss between S1 and S2. Mania is related to metonymy, automatic in nature, to such an extent that it erases the desiring subject. In the last meeting of *Seminar 10*, Lacan says that in mania it is about the absence of function of the object *a*, and no longer simply its ignorance. Not only is the object of jouissance unknown, but it is absent. The subject is no longer weighed down by any object, it is not ordered by an axis, but rather it is about delivering that object to infinite metonymy with no possibility of escape. In this way, mania does not present a state of subjective possibility, rather it is the effect of the invasion of jouissance to the point of eliding all possibility of desire that results in a dead subject as abolished in its relationship with desire.

To summarize, Bazan and Detandt dynamically locate jouissance in the motor tension of those action programs marked by the mesolimbic mesocortical dopaminergic system, that is, they locate jouissance in signifiers associated with reward seeking. Dall'Aglio confirms this model by postulating a "motoric" unconscious, which makes the disjunction between symbolic and real disappear. The signifier is given a neural basis as a motor articulatory pattern. Between the dopamine peak and the appropriate action to satisfy the need, says Dall'Aglio (2020b), there is a weld. "Therefore, it can be postulated that animals have some kind of jouissance, another point where neuropsychoanalysis can challenge Lacanian theory" (p. 732). However, for Lacan (2008f) "[…] the unconscious has not revealed anything to us about the physiology of the nervous system […]" (p. 139).

Life outside of the being that speaks

The neuroscientific studies that are done regarding the mind are, for the most part, on the brains of Bonobo or Rhesus monkeys because they are considered the closest animals to the human being. In these studies, the scientist avoids having to listen to the patient because if the central issue is in the neurons, what is the use of free association? In this sense, Kandel (2009a) affirms that psychoanalysis has not evolved scientifically because it has not developed objective empirical methods used in non-speaking animals to test its hypotheses, which made it obsolete. He wants to bring psychoanalysis under the halo of an objective science that reduces the clinic "to a quantification of biological and genetic markers, without the need to resort to the thorny and confusing testimony of the patient's word […]" (Bassols, cit. Zack, 2016, p. 11).

The neurobiologist looks for objective criteria, otherwise they are subjective, that is, not scientific. In his conception, talking to someone can be nice, but the speech cannot be identified "inside" the brain. Freud created a dialogical device, so access to the unconscious is not through ways that do without the subject speaking, but rather it is to the extent that there is someone who knows how to question him.

The conception of life outside of the being that speaks implies elevating adaptation to the central axis. In this sense, the story that Kafka (2011) addresses to the academy about how a primate has managed to access the human world thanks to the education it received from the environment that civilized its sexual and aggressive instincts is well known. But let's see what happens:

> In the story the monkey is found in the jungle and brought to the life of the University. He attends the academy to testify to his monitude, and his passage from monkey to man [...] the first thing that learns is to spit, because all the sailors who bring him on the ship, caged, spit, so he spits too; the monkey's first human sign is spitting.
>
> (Chamorro, 2011, p. 54)

So adapting means not criticizing. This skips the question of whether the word is empty or full. If psychotherapy, Kandel argues, works at the level of the synapses, then there is no need to speak. For example, dyslexia has been associated with an abnormality in the brain, so it is no longer the exclusive property of humans, the mouse can also be dyslexic. In this way, various authors (Judith and Rapoport, 2009; Kandel, 2009b) affirm that psychoanalysis should be aligned with psychiatry to approach subjects from the psychopharmacological model. What comes is the ideal of being able to diagnose and treat a disorder without having to exchange a single word with the patient (Bassols, 2014). There is no direction to a subject supposed to be able to produce the missing meaning, the key to an enigma. The subject approached from the psychopharmacological model produces its call to the representative of the universal, such that it can erase all singular joussicance in the ocean of statistics. The subject approached from the psychopharmacological model expects its certainty to be authenticated (De Georges, 2005).

In this sense Liberman (cit. Bazan, 2011) proposes the motor theory of speech perception, based on phonological research, which "holds that the basis of speech perception is not the actual sound of speech, but the articulatory gestures made by the speaker" (p. 168). They have found a brain area, which they call "F5", which is activated when articulatory gestures of speech are registered, which stimulates Broca's area. This area would be made up of mirror neurons in the human speech motor areas of the brain and with this they explain why lip reading improves the intelligibility of what a person says. This area has also been found in the monkey: "Of particular importance is the fact that [...] the F5 area in the monkey is the probable homologue of Broca's area in humans" (p. 169). In this sense, Corballis (cit. Bazan, 2011) affirms "that the origins of human language could be located in the manual gesture rather than in vocalization" (p. 169). Conclusion: it is not necessary to speak, moving your lips or hands is enough to understand.

For his part, Kandel (2009b) has found that the increase in size of INAH-3, the most important of the sexually dimorphic nuclei of the hypothalamus, is related to male heterosexuality. Likewise, various authors (Gorski, 1996, cit. Kandel, 2009b; Scvhiavi et al., 1988, cit. Kandel, 2009b) have discovered that electrical stimulation in regions of the hypothalamus causes typical sexual behavior in rats and rhesus monkeys, so they are sure to affirm that sexual behavior is due to the neuronal distribution of a region of the brain. Schiavi et al. (cit. Kandel, 2009b) conclude that "[...] sexual orientation depends on genes" (p. 96). The authors believe that by studying these issues in simple organisms, such as snails, and slightly more complex ones, such as mice and monkeys, conclusions could be extended to human sexuality.

> An important and almost always unconscious cerebral nervous mechanism synchronizes the movements and in particular the courtship dances of two or more participants. Recently discovered, such a mechanism is based on mirror neurons, discovered in rats and monkeys, although they also appear to be found in numerous animals, including humans.
>
> (Langaney, 2006, p. 82)

Would it be saying that sexual desire is chemical? Jean-Pierre Changeux (cit. Castanet, 2023), a globally important neuroscientist, affirms: "The orgasm is, therefore, above all a cerebral experience, and it is in the brain that we must look for its trace" (p. 86). For his part, Otto Kernberg, once president of the IPA, has argued that sexual experience rests on the biological foundations of the bodily apparatus of jouissance (Kernberg, 1998). "According to Kernberg, desire can then go from division to unification [...] It is about giving a biological foundation to subjective division" (Laurent, 2002, p. 84).

This is what English-speaking countries traditionally prefer, in which, unfortunately and this is something to change, the teaching of psychoanalysis is not promoted, or yes, but in its closed form, they want very defined, very strict concepts; they want usable knowledge, with the exception of a few psychoanalysts in relation to science, they do not want to open discussion, they do not want the Freudian Field. Regarding this, Miller (2010b) points out that Otto Kernberg, for example, said that he was very worried because he could not grasp the exact definition of Lacanian concepts. "They change all the time," he said. We can imagine dear Otto, and the aforementioned neuroscientists, wanting to find in Lacan the definition of the unconscious, of the Name-of-the-Father, of the signifier, which is not found with one, but with a plurality of definitions. They find themselves with contradictory definitions and remain just as lost with Lacan. The unification statement can be translated into a maxim: nothing is impossible. There are no obstacles to the aspiration of integration that gives the ultimate meaning by erasing the boundaries of the subject in a confusion with the soma. Kernberg manages to save the ideal of the sexual relationship "and keep psychoanalysis in the range of virtuous and effective disciplines that are part of what can improve the functioning of each one" (Laurent, 2002, p. 125). Kernberg (1998) characterizes desire as the identification of arousal with sexuality, as a complementary experience of fusion. Even sexual orientation is encoded in synapses:

> With the usual brain research methods, it is already possible to obtain information about the personality of a subject, about possible risks to his health or his inclination

to aggressiveness. Sensitivity, dependability, pessimism, risk taking, extraversion and neuroticism, as well as sexual orientation and unconscious ethnic prejudices, are some examples of psychological traits, which can be assigned to certain characteristics of brain activity. Without forgetting the incessant search for lie detectors, each time better and more reliable, in these times of war against terrorism.

(Langaney, 2006, pp. 90–91)

These ambitious interpretations are traversed by the ghost of the person of the scientist. While the training of the psychoanalyst implies that, with control, his analysis, and the relationship with the school, he can clear the analyst's desire function of his own symptoms and abstain from his own fantasies in order to conduct the cure according to the significant function that is appropriate in each subject. "The training of the analyst means that the analyst has the broadest possible position, the least determined by his prejudices, judgments, and that he works protected by the patient's discourse [...]" (Chamorro, 2011, p. 143).

In these discoveries, the neuroscientist intervenes with his judgments, covered by the fact that they are "scientific judgments". Taking the floor quickly, he listens to two or three signs and goes ahead, he already knows the diagnosis, he knows what is coming. It anticipates, structuring a deceptive transfer, in the form of a supposed knowledge that does not correspond to the consultant, but to the knowledge of science, that is, to nominalism. In his briefcase is the DSM (*Diagnostics and Statistics Manual*) that says the criteria to circumscribe the subject and from there the treatment comes attached. It is like that for everyone, it is proven, that it is effective. It is the replacement of the statistical reason by the logical reason. Freud advised against this mode of operation: "As I have found out, there are analysts who boast of instant diagnoses and treatments on the run, but I warn everyone that these examples should not be followed" (Freud, 2012k, p. 141).

There are analysts, such as Wallerstein (2004), president of the IPA at the end of the 1980s, who affirm that what psychotherapy and psychoanalysis share is the cure by convergent criteria. What matters is that the treatment leads to an answer that is not ambiguous. "The treatment is not tailored to each subject, but it is entering the universal of the structure" (Chamorro, 2011, p. 65). In this perspective, the basis is the diagnostic criteria that identify the subject with some dominant signifier or set of similar people. The problem is that where there are criteria, there are no questions and if there are no questions, there is no way for a subject to contribute what is unique about it.

Even the foundation of sleep, suggest Born and Wagner (2006), is located in the neural network. Then Provine (2006) affirms that yawning reveals the neurological basis of empathy and unconscious behavior and Solms, when asked by Steve Ayan what relationship exists between sleep and unconscious conflicts?, maintains: "Jaak Panksepp, from Bowling Green State University, discovered years ago in experiments with rats a 'search system' in the back of the brain, the called the ventral tegmental area" (Ayan, 2006, p. 75). All of which means dispensing with the subject saying something about his own dream.

It is true that with neurobiology research resources appeared that led to other more effective treatment practices, neuroimaging allows us to see brain function without opening the skull, and modern technologies record glucose or oxygen consumption and thus see which brain areas present abnormalities. But there are also abuses, the rapid and automatic extrapolations that are made from animal model research in artificial environments to human conditions.

> Due to his profession, the biologist believes in the sexual relationship because he can base it scientifically, but at a level that does not imply that it rests on the unconscious [...] Even when the biologist verifies the way in which the sexes relate to each other, he does so at a level where this does not speak.
>
> (Miller, 2010, p. 52)

So, neurobiology is prescribed for the sexual relationship, and this is the root of its potency (Laurent, 2002). It is the reduction company that says it follows Freud, but generates the loss of its own logic, that is, its abolition. Solms (2006) himself says it when asked what would be different in psychoanalysis today if Sigmund Freud were alive? he replies, "I'm pretty sure that practice wouldn't exist" (p. 75).

Cognitive machine

The previous notion, which cannot avoid that one of its consequences is to omit the subject's word, gives rise to a therapeutic method based on the interpretation that the subject handles information like a machine. According to Kandel, psychoanalysis did not evolve because it was not developed as a theory of information in the brain: "Although Freud's ideas were very interesting and had an influence, many scientists were not convinced by them because there was no experimental investigation into the information storage system in the brain" (2007, p. 161). That is why Pinker, a Harvard professor, investigated Freud to defend the thesis that the mind works like a system of physical engineering organs.

> [...] It is not difficult to find explanations such as, for example, victims break down under pressure, children are conditioned to do this or that, women are brainwashed into valuing this or that, girls are taught to be such and such. Where do these explanations come from? Undoubtedly, from the hydraulic model of the mind developed in the 19th century and picked up by Freud.
>
> (Pinker, 2001, pp. 84–85)

Pinker says that "the hydraulic model, with its cumulative psychic pressure and discharge of energy, its explosions or its diversion into alternative channels, is at the center of Freudian theory" (pp. 96–97). He interprets a hydraulic model in Freud "in which psychic pressure develops in the mind and can explode unless it is channeled into the appropriate paths" (2004, p. 145). Then he says that this is false, that the mind does not work by flow of energy, but by information. So, Pinker makes Freud go from engineering theory to computational theory, a passage that includes evolutionary notions, where information is processed by the organ of perfection, the brain, and the way in which this processing occurs would result in the survival of the being in the environment: "the key to understanding the mind is to try to perform its reverse engineering, to calculate what natural selection designed it to adapt to the environment in which we evolved" (p. 145). Likewise, Talvitie (2009) compares the unconscious with a system that stores and returns information. He is convinced that the unconscious can be read from the theories of Turing, Fodor, Stich, Churchlands, Dennett, Nyman, among others.

> There is a brutal irony in the life of the subject Alan Turing that it is convenient to summarize it here. After having discovered and cracked the "Enigma" code of the German navy, the British government wanted to "reward" Alan Turing with treatment to correct his homosexuality. In reality, he was first prosecuted for his homosexual orientation, and was given a choice between jail or a reductionist treatment: the so-called chemical castration. The idea, so scientific, was to correct a supposed programming error in his body with an injection of hormones. Alan Turing chose to continue the treatment. Tortured by its various consequences, he died eating a poisoned apple. It was officially ruled a suicide. There are reasons to think that his destiny would have been different if he had been listened to, not only from a non-moralist position, but above all from a non-reductionist orientation of the subject. We see how subject Alan Turing was himself treated as a Turing Machine.
>
> (Bassols, 2011a, p. 81)

This brutal irony highlights the possible consequences of interpreting the subjective from a programming theory. For Talvitie (2009) psychoanalysis is a theory extracted from the same logic with which machines are thought: exact proportions and fixed relationships between elements. "Issues such as neural algorithms, 'calculations' and neural networks, or the brain as a system, seem to work in a 'biased' way, preventing certain ideas from emerging into consciousness. With this theme we deal with issues that Freud addressed [...]" (p. 13). So, Talvitie tries to give the unconscious an ontology that Freud did not know how to give:

> The previous idea that the repressed contents are only neural representations whose activation is inhibited is easy to understand when thinking about a computer. A computer contains digital representations of, among other things, the images you have taken with your digital camera. Clicking on an appropriate icon activates the digital representation of a certain image. Paraphrasing Freudian thought, we could say that by clicking on the icon, the image is taken or transformed from the computer hardware (the brain, the unconscious) to the scope of the screen (consciousness). We may think that representation can also be prevented from entering the domain of screen/consciousness.
>
> (p. 113)

When clicking, the knowledge in the pocket, which implies a short distance to access it, and is no longer the knowledge as an object of the Other (Grinbaum, 2022). This indicates that the new platform of the social bond is digital life. As Byung-Chul Han (2022), a South Korean philosopher based in Germany and professor at the University of Berlin, says, digitization is an idea that comes from *digitus*, a Latin word meaning finger. In digital, human action is reduced to the tips of the fingers, it is only within reach of a click. We have facilitated the exploitation, the investment of money, the stripping, the exposure in networks, and the diagnosis by images. Today we only move our fingers, "it is the digital lightness of being" (p. 133).

The machine is encrypted, it is a fixed program like the human genome, it has to close, it cannot disagree with itself, it cannot be wrong, it defends the agreement between its own elements, while the unconscious is equivocal. Hence, Talvitie has questioned the Freudian concept of the unconscious: "Is Freud's perspective on the unconscious, the cornerstone of psychoanalysis, correct? [...] Was Freud right after all?" (2009, p. 15). Likewise, Buchheim, Cierpka, Kächele and others (2013) point out that Freud appreciated that the state of science at that time would not allow explaining the psychic transformations of the brain and that thanks to the neurosciences of the late 20th century that began to unravel the mechanisms of unconscious information processing through neuroimaging techniques "Freud's longed-for vision was fulfilled, one hundred years late" (p. 27).

The dream of turning man into a machine is a fact. CONICET has projected a mobile application for technological devices designed to help health science professionals diagnose schizophrenia through the analysis of patients' discourse. This application records a speech, automatically analyzes it and detects, based on the speech patterns – number of verbs used by the speaker and discursive coherence – the probability of suffering from schizophrenia. Slezak (2018), one of its developers, says: "What we did with these interviews was to develop an automatic analysis of the texts and quantify the messages through certain characteristics, and predict which high-risk patients would trigger schizophrenia" (p. 2). Now, if psychotherapy is going to be based on robotic codes then psychoanalysis, that is, having someone listen to you once or twice a week, will be a luxury.

However, the advance of science and the market, Peteiro points out (cit. Bassols, 2011a), reveals a depreciation of the authority of doctors or neurologists by those who need them, which goes hand in hand with the increase in credibility placed on technical devices. If you have an appliance you have power. Even neuroscientists themselves "have been fascinated by the power of the technique" (p. 202). The medical act is becoming dehumanized, it is being diluted in the knowledge of science, which is no longer given by the person of the doctor thanks to his art of listening, looking, auscultating or interpreting, but now depends on the machine. The neuroscientist himself "is becoming an intermediary, a technician who will be replaceable by a robot [...]" (p. 202).

What do we have on the horizon? At least a double threat. On the one hand, a strategy of standardization of psychic problems leads to psi intervention through robotics. As Han (2022) says, we live under a form of capitalism that preaches immediacy, waste, and the dream of eternity, where the imperative of jouissance pushes mere moments without time to understand. Living fast means not getting entangled with someone's life story, which undermines the possibility that someone could enter a clinical device. Benjamin (1982) stated that the exchange of experiences is being lost due to its fall in monetary value and that the art of storytelling is nearing its end, while it is being replaced by a new form of communication: information. On the other hand, the elimination of trust and the Other, because one speaks to the bug and the artificial intelligence answers. If you tell him that you feel good, the pager sets, for example, that you are in the park; when you tell him that you feel bad, the bug will send you to the park! It is about receiving advice at a low cost.

Encrypted data and automatic response. Now, there is talk of neuroplasticity, of change, but the encryption, the fixity, is maintained. How do they solve? With neurolinguistic programming, which arose in the 1970s from the investigations of Bandler and Grinder (1980; cit. Barros, 2004). The point of view of this current is based on the conceptual model proposed by Noam Chomsky, which postulates the linguistic nature of the processes that give rise to the construction and functioning of the psychic apparatus from a dynamic perspective of information, where the terms measurement, union, quantity and accumulation are fundamental.

In this approach, knowing is thinking. Although the particularity of a machine is that it does not speak, it can think. The large amount of data storage and the connection between them at a speed impossible for humans gives the impression of thought. "Lacan points out that he is willing to consider the idea that a machine thinks – which is already quite a gamble – but we do not have any evidence that a machine knows anything. To know, nothing at all" (Bassols, 2011a, p. 80, 2016b). For Lacan, handling information is not knowing. Information and unconscious knowledge are two very different things (Bassols, 2013b). The construction of a database with a quantity of information does not imply that a subject can find meaning in that quantity. As Bassols (2012b) says, to know you have to enter into dialogue. You can't talk to a machine. A machine cannot give signs that it is enjoying itself.

> They see that the problem is more complex than the being that speaks as a cybernetic device supposes. The being who speaks, who is submerged in the field of language, does not function from the traces of something that was inscribed in memory, but from traces erased by language itself.
>
> (2011a, p. 98)

From Lacanian-oriented psychoanalysis, Laurent (2010) points out that the analyst has to help the psychotic subject to stop the thinking machine. The psychotic subject will not stop interpreting, on the contrary, he suffers from not having punctuation, it is one interpretation after another, that he analyzes without stopping. It is not the psychotic subject who will stop interpreting, but the analyst has to help the subject to name that unnameable, to introduce a pause in the unlimited jouissance. The analyst must wait for that. Holding the translation of the psychotic subject while waiting for a name to be made. Making a name for himself allows the psychotic subject to stop this endless process. Interpretation in psychosis is the moment when the system stops. We aim to obtain a stopping effect, which ceases the infernal machinery of non-stop interpreting. A moment of "thinking about nothing", of stopping, can allow the subject to sustain himself without being hospitalized. It does not mean silencing the psychotic subject, but obtaining a pause, that is, a score in the interpretation of the psychotic subject, something that is not possible in neuroscientific theories where there is more and more knowledge. It is the analysts are the ones who act as the *capitoné* point, the ones who try to obtain pauses (Laurent, 2010).

The accumulation of data drives you crazy, and feeds the dream of calculating what the other wants (Laurent, 2011). A kind of "bring the brain you have and take the brain you want". The much-contested lobotomy at the service of the promising intervention on the brain. The subject becomes an element of the machine, like a car: when something doesn't work, you have to look for proportional replacement parts (Laurent, 2007). Proportion is the basis of this notion. On the other hand, the training of the psychoanalyst does not imply a quantity of information (Balzarini, 2021). The knowledge that is involved in the unconscious has nothing to do with information, memory storage, learning, or pedagogy, but rather a knowledge that is housed in the "discourse in which the unconscious is interrogated in the manner of saying why!" (Miller, 2015a, p. 188).

Already in the 18th century, a book entitled *Machine Man* suggested that the human being could be thought of as an automaton. This is the dream of current scientism, which returns with a new instrument, MRI – magnetic resonance imaging – an essential tool for its research (Bassols, 2011b). Indeed, the quantity sciences have exploited these instruments, with which they can determine unconscious information. If we take this to the extreme, by transferring gray matter we export the sexual position of one being to another. The hardware would be the brain and the software would be the unconscious. Software is reproduced on numerous hardware, experiences from one being to another. It is in the imagination of scientists that "the brain can benefit from the accumulation and cultural transmission that spanned millennia" (Miller, 2015a, p. 184).

One can capture how the neurosciences, and their cognitivist partners, try to establish themselves in an ideological perspective that tries to maintain that the human being is comparable not only to the rats that are used for experimentation, but also to the machines in their way of processing the stimuli to which they are subjected (Zack, 2008). The consequences can be verified, it is what Han (2012) calls the fatigue society.

Han's thesis is that the subject exploits himself because growth in capitalism goes by way of accumulation. Accumulation means an eternity of being, a non-finite being. Accumulation provides the illusion of a life where death is not a part, an undead life. Charles Melman (a student of Lacan), said that "Lacan's way of proceeding was a way of pointing out that we are not eternal" (cit. Rosales, 2017, p. 66). We think we are eternal, but we are not. There is urgency, it is necessary to hurry, leave nothing for tomorrow, it has to be now, don't wait. "Negativities such as pain are eliminated in favor of the possibility of satisfaction" (Han, 2022, p. 27). What is different is left aside to speed up the cycles of production and consumption. The otherness is taken away from the Other and there "we can no longer love him, but only consume" (p. 129). An excess of positivity is what dominates the current social phenomenon. The drive to perform at your best. You have to enjoy, not ask. If a child says that they want to change their name, they are automatically offered a battery of possibilities between hormonal and surgical interventions. The market offers this push to enjoy. The violence then does not come from the negative, but from this extreme of positivity (Han, 2022).

What happens in medical centers or in psychopathology services is that whoever aspires to be a patient is no longer concerned with knowing where the therapist comes from, if he is trained or not, but only if he is useful. "One is interested in knowing if that works, here and now" (Miller, 2004b, p. 11). It is as Deng Hsiao-Ping (cit. Miller, 2004b) said, who made the Chinese New Testament: "The color of the cat does not matter as long as it catches mice" (p. 11).

The prevalence of the ideal of control toward productive ends produces a subject of pure competition, that is, without the unconscious. Hence the creation of so many business schools that teach the background of a successful businessman, with the computer as an instrument of great human achievement. And no one protests about this, on the contrary. In the capital society described by Marx, the worker was exploited, but after a certain level of production it reached its limit: protests. On the other hand, in the neoliberal society that Han describes, the oppressive instance is apersonal, there is no someone or something against which the subject directs his fighting force. A "we" is not constituted, a collective that can rise up against the system is not erected, it is the silence of the death drive. A non-repressive system, rather tempting. In it, the human being is the free producer, isolated, "self-taught", exploiter of his own self. Today everyone "is lord and servant in the same person" (Han, 2022, p. 33). Which provides the sensation of freedom, but it is paradoxical, since it is nothing more than slavery at the service of performance.

> The beginning of the 21st century, from a pathological point of view, would be neither bacterial nor viral, but neuronal. Neuronal diseases such as depression, attention deficit hyperactivity disorder (ADHD), borderline personality disorder (BPD) or occupational burnout syndrome (OSS) define the pathological landscape at the beginning of this century.
>
> (Han, 2012, p. 11)

What Han says is that every age has its emblematic diseases. At this time, diseases involve an excess of positivity, we have it in hyperactive children, "which has its origin in the overabundance of the identical" (p. 23). All stimulated by the same instructions, there is no mediation with the Other, which produces psychic illnesses, such as brain burn syndrome, which constitute the pathological manifestations of a paradoxical freedom.

The subject subjected to the pressure to produce never reaches an end point. One permanently lives with a feeling of lack that is maximized by the virulence with which the feeling of guilt is entrenched. We no longer compete with others, but with ourselves. It is the thesis of Alain Ehrenberg (cit. Han, 2022), that if there are so many depressions in this world, it is because the reference to the conflict has been lost. Hence, the neural subject has two paths left: success or failure. In that it is similar to machines: it works or it does not work.

The difficulty is that they do not produce responses from the body, such as fatigue and exhaustion, that are "manifestations of neuronal violence [...]" (Han, 2012, pp. 19–20). The increasing prescription of antidepressants in the world goes hand in hand with this drive to erase all conflict and produces the rise of self-affirmation techniques that prepare the subject to face stressful situations. They are techniques that work as long as the subjects repeat them insistently to give themselves security: I have to reaffirm my rights, I have to have a positive image of myself, fight depression through action, I have to be respected. The affirmation of oneself makes the subject the master of himself. "Any critical sense seems to be abolished with the only perspective of changing man to maximize his ability to learn" (Simonet, 2019, p. 9).

In short, the diseases of this time assume "that the reduction of human reality to the brain is legitimate, that man is essentially a brain and that the brain is an information processing machine" (Miller, 2015a, p. 145). Now, if the subject breaks down into a series of microcerebral treatments from blood flows, then the subject is no longer right (Lardjane, 2019).

The id, affective consciousness

Mark Solms (2013) has the thesis that the fundamental matter, the essence of consciousness, is affect, and not consciousness linked to perception (conscious vision, conscious hearing), which is the way he says that Freud conceived consciousness. According to Dall'Aglio (2021), affective awareness cannot be reduced to cognition, it is only affectivity; it is the mere connection with the body. Thus, affective consciousness is, according to Dall'Aglio and Solms, the nucleus of the id. What Dall'Aglio suggests is that the material of primary repression is not ideational, as Freud argues in "The unconscious", but raw affect without an object. Solms and Turnbull (2002) add that what is not represented, what does not reach declarative memory, is also linked to cognition. In this sense, an affect can be completely said and exhausted with words. The development of the hippocampus, in charge of depositing declarative and episodic memories, allows these capacities to encode the objects of affection. Memories can be recovered in working memory thanks to the fact that any object can be encoded in these affective systems, which constitutes, according to the authors, the unconscious ideas proposed by Freud in the text "The unconscious". With these considerations, the authors propose a change in the Freudian notion of the unconscious. Instead of an anti-cathexis that keeps the repressed primary inaccessible, they propose that the unconscious is the repository of memories stored by objects encoded by affective systems based on lived experiences that can be translated into representations that can reach cognition through the development of the hippocampus, which gives the ability to name objects. "In this way, the distinction between affective consciousness (raw affect) and cognitive consciousness (representation capacities) reviews Freud's unconscious" (Dall'Aglio, 2021, p. 140).

They trust so much in the recovery of the repressed that they call the core of the unconscious "affective consciousness", that is to say, what is in the core, the depth of the unconscious, is designated with the word "consciousness". There is nothing that cannot be represented if and only if the hippocampus achieves further development. And what cannot be represented, due to failures in the development of the hippocampus or because the hippocampus has not yet developed, constitutes what is called the motorized unconscious, automated habits. So we have affective consciousness at the base as the first level. Then, the affect, the rawness of the feeling, joins a second level, the cognitive conscience, by connecting the drives to procedures. Finally, the subject can be self-aware of what they feel at a third, more elaborate level called tertiary consciousness. In this way, "a neuroanatomical basis for meaning is provided" (Dall'Aglio, 2021, p. 142).

The neurochemical basis of libido

We know that the greatest developments of the neurosciences have been translated into English. However, a translation always betrays the original author. Sometimes you read things that are strongly confirmed when they were not exactly said that way by the author. For example, the translation of Freud's work from German to English has paired the terms *trieb* and drive, a pair that was translated into Spanish as impulse. The reality is that in "Drives and drive destinations" Freud chooses the word *trieb*, which in German means drive, instead of the word *instinkt*, which exists in German and means instinct. If Freud had wanted to say instinct he would have written *instinkt*. The translation of the Amorrortu edition respects the word that Freud chooses with great precision, which is *trieb*, but in the Biblioteca Nueva edition it reads instinct.

Translations facilitate the transfer term by term, in this case the passage to chemical language: "for Freud this new method was not fundamentally different from the microscope, in terms of its scientific objectives" (Solms and Gamwell, 2006, p. 17). Likewise, Solms and Solms (2005) understand that Freud sought to provide a chemical basis for making inferences about the underlying functions of the unconscious that could not be directly observed.

> The libido, and especially its narrowly sexual components, is accessible (in principle) to chemical means of analysis. It is firmly rooted in the physical processes of certain body tissues [...] This makes it possible for us to link the destructive instinct, in turn, with a more primitive physiological tendency of the nervous tissue.
>
> (Solms and Solms, 2005, pp. 289–290)

In this sense, Solms (2007) states that Freud established the concept of libido identical to the reward system mediated by the dopamine agent. And Delgado, Strawn, and Pedapati (2015) state the following:

> Freud [...] proposed that, from a biological point of view, an "instinct" appears to us as a concept on the border between the mental and the somatic, as the psychic representation of the stimuli that originate within the organism and reach the mind.
>
> (p. 123)

The identification of a chemical basis for the concept of drive brings it closer to that of instinct. Precisely, Ansermet and Magistretti (2006) say: "[...] the psychoanalytic concepts of the unconscious and the drive acquire a biological resonance [...] The question of the biological status of the unconscious and of the drive is, then, raised" (p. 218). And Kandel states:

> In studies of the neurobiology of emotional behavior, David Anderson of the California Institute of Technology has found some of the biological underpinnings of two of the drives Freud observed, eroticism and aggression, as well as the fusion of those drives.
>
> (2018, p. 364)

That is, we have a structure where the drugs must be sent to control the conditions that produce eroticism. In fact, Solms and Turnbull (2011) have located the chemical basis of the Freudian concept of drive. They illustrate it with some quotes from Freud, such as one from "Introduction to narcissism":

> We must remember that our tentative ideas in psychology will presumably one day be based on an organic substructure. We are taking this probability into account by replacing special chemical substances with special psychic forces.
>
> (Freud; cit. Solms and Turnbull, 2011, p. 134)

From *Beyond the Pleasure Principle* they take this quote:

> The deficiencies in our description would probably disappear if we were already in a position to replace psychological terms with physiological or chemical terms [...] Biology is truly a land of limitless possibilities. We can expect him to give us the most amazing information, and we cannot guess what answers he will return in a few dozen years to the questions we have asked him.
>
> (Freud; cit. Solms and Turnbull, 2011, p. 134; cit. Solms, 2015, p. 15)

If it is read as it should be, Freud is not supporting the chemical and drive relationship, quite the contrary. He says that if this relationship were legitimate there would be no deficiencies in the descriptions. If there were a relationship, there would be no deficiencies and the drive could be controlled. It is the idea of Edelman and Tononi (2002) who affirm that learning is at the level of the drive. Radically opposed to Freud (2011b), who indicated that the drive is always not all and that education is intrinsically linked to the impossible. In this way, in 1914, he proposed the first drive dualism, the conceptual division between sexual drives (those that aim to seek the object outside of the self) and self-preservation drives (*yoíc*) for which he provides three arguments. First, he says that hunger and love are two totally different things. That is, it separates the biological from the psychic. Second, it says that there is a force against the pleasure principle, which endangers survival and is capable of breaking the equilibrium, its power lies precisely in the fact that it is not a biological property. Such the second argument:

The individual actually leads a double existence, insofar as he is an end to himself and a link in a chain to which he is a tributary against his will or, at least, without his will. He has sexuality for one of his purposes, while another consideration shows it as a mere appendage of his germinal plasm, at whose disposal he puts his forces in exchange for a reward of pleasure; he is the mortal bearer of a – perhaps – immortal substance, as an ancestor is only the temporary beneficiary of an institution that survives him. The separation of the sexual drives from the ego drives would only reflect this double function of the individual.

(Freud, 2012ll, p. 76)

And third, he says that he should rely on biology to separate both drives, but he takes this as a probability:

Thirdly, it must be remembered that all our psychological provisionalities must eventually settle on the soil of organic substrata. It is probable, then, that particular chemical materials and processes are the ones that exert the effects of sexuality and act as intermediaries in the continuation of individual life in the life of the species.

(p. 76)

He says, "It is probable". Six years later he says again:

It is probable that the defects of our description would disappear if instead of the psychological terms we could already use the physiological or chemical ones. But in truth these also belong to a figurative language, although it has been familiar to us for a longer time and is perhaps simpler.

(2012h, p. 58)

To say that it is probable is to say that it is not a fact, hence Freud spoke of the representative agent. Why is Freud made to say that the probable supersedes the true? The intention of these authors is then to say that the neurosciences prove Freud (Laurent, 2020b). Says Solms and Turnbull (2011): "Freud, in our opinion, would have considered this a welcome and entirely legitimate development of the work that he pioneered" (p. 134). Solms also takes this paragraph:

On the other hand, let us note well that the uncertainty of our speculation was greatly increased by the need to borrow from biological science. Biology is truly a realm of limitless possibilities; we have to wait for the most surprising clarifications from her and we cannot envision the answers that decades later she will give to the questions that we put to her. Perhaps he will give them such that they will collapse our entire artificial edifice of hypotheses. But if so, you might ask yourself: Why take jobs like the ones listed in this section, and why communicate them at the same time? Well then, it's just that I can't deny that some of the analogies, links and links pointed out in it seemed worthy of consideration.

(2012h, pp. 58–59)

We can expect biology to provide explanations of certain structures from which chemical processes necessary for the continuation of the species are derived, but in the human field it would only provide probabilities. And if biology gave these missing explanations, says Freud, it would collapse the hypothesis that gives rise to psychoanalysis: the hypothesis of the unconscious. Freud is clear that biology has nothing to do with psychoanalysis. "Precisely because I have always strived to keep away from psychology everything that is alien to it, including biological thought [...]" (Freud, 2012ll, p. 76).

There is no intracerebral relationship

Dall'Aglio (2020a) questions the emphasis that Lacan placed on the insubstantial subject of the unconscious, since he understands that this emphasis makes the unconscious seem radically incompatible with the brain. Hence, Dall'Aglio criticizes the position of Lacanian psychoanalysts that the brain does not know the drive. He recognizes that the subject of the unconscious cannot be reduced to biology, so in order to support Lacanian neuropsychoanalysis what he does is show that the brain is not reduced to biology. To do this, he distinguishes neurodiscourse and the study of the brain. The first is the all-encompassing promise of knowing, which rejects impossibility and cannot withstand the critiques of psychoanalysis. But the study of the brain, says the author, citing Aguiar (2018), opens lines of research that show the increasingly indeterminate real and open to modification, that is, a real without law and with it the passport to Lacanian neuropsychoanalysis. In addition, he says that Lacanian psychoanalysis will benefit from being combined with neuroscience because it will have a better capacity to participate in the mental health discourse. If psychoanalysis is combined with neuroscience, it will adapt to the prevailing discourse and it will be a therapy like the others, that is, a normal therapy. And this would go against Lacan.

Dall'Aglio (2020a) cites the study by Copjec (2015) from which he recognizes that the subject is what escapes neuroscientific quantification, but this does not mean, say the authors, that the subject has no relationship with the brain. The real is not an impossibility located at an external limit where only psychoanalysis can operate. The real is an internal limit, an internal impossibility. To illustrate the possibility of a Lacanian neuropsychoanalysis, Dall'Aglio draws specifically on the philosophical work of Adrian Johnston, which we have already presented, who, taking the Lacanian style, states: "there is no intracerebral relationship" (Johnston, 2013, p. 59).

Johnston proposes that the brain itself must be understood as divided through the recognition of a neural non-relationship. He emphasizes the division between non-representational subcortical affective structures (affect-related brain structures that are unrepresentative) and neocortical declarative structures (responsible for language and reflective cognition). The idea of Johnston (2013) and Dall'Aglio (2019; 2020b) is that nature is divided from the beginning, there is a structural negativity manifested as discordant relationships between these brain systems. Under this negativity, a surplus is produced where the real is established in the brain. In this way, Johnston reads Lacan's thesis, equating it with his own thesis "There is no intracerebral relationship". Everything is inside the brain, the impossibility is armored there. They support intracerebral rupture and, at the same time, neural connectivity, which is a paradox.

RSI brain mapping

Dall'Aglio (2019) locates the three Lacanian registers in dynamic parts of the functioning of the brain, which avoids neurostructural reduction, that is, it seeks to advance one more step from the metapsychological bridges that Solms and Solms (2005) had established between neuroscience and psychoanalysis when they had focused on Freud's structural model. If Lacanian registers are dynamically located, a particular organization is reflected between constellations of neuronal activity, that is, it prevents the reduction of a concept to a single neuronal explanation.

For this, Dall'Aglio takes the model of Johnston (2013), who affirms that "the cerebral cortex, through the thalamus, is the conduit through which the phenomena and structures of the Imaginary–Symbolic reality affect and mediate the corporal Real, embodied above all by the brainstem" (p. 62). According to Johnston, the real corresponds to non-representational emotional-motivational subcortical systems, while the symbolic and imaginary correspond to neocortical representational systems. What Dall'Aglio contributes is that the right and left cortical hemispheres seem necessary, but not sufficient for imaginary and symbolic functions respectively; that the real is underpinned by subcortical affective and motivational structures, particularly the automation of non-representational systems before the development of the hippocampus. So Dall'Aglio's intention is not to reduce Lacanian theory to neurobiology, but to elevate neuropsychoanalysis beyond neurobiology, as it was, according to him, the intention of Lacan.

Regarding the imaginary register, he indicates that "the right hemisphere seems to be a key place in the brain circuits that underlie the constitution of the ego as a gestalt, thus serving as the basis for the imaginary register" (2019, p. 34). Through such a hemisphere the partial impulses are unified and the image is not torn to pieces. Likewise, Vanheule (2011) affirms that Lacan's hypothesis is that the mirror stage is located in the cortical area of the brain because it has the function of uniting fragmented partial impulses and repairing the rupture in the relationship between the object and the ego. This idea, according to the author, supports the connection that Lacan makes in *Seminar 1* between the imaginary internalization of the ego and the constitution of external perceptual reality. With which he affirms that the damage of the right hemisphere distorts the representation of the body image and is what causes various cognitive pathologies.

Regarding the symbolic register, Dall'Aglio (2019) suggests "a linguistic organization of signifiers in the brain connected to semantic meaning" (p. 33). If the brain has symbolic areas that "underlie meaning" (p. 33) then, as various studies have identified (Binder and Desai, 2011; Price et al., 1997; Rapcsak et al., 2009), the left hemisphere is the seat of language. Damage to this hemisphere produces deficiencies in producing abstract links between concepts, for example, aphasia or language disorders. In this sense, McGilchrist (2009) indicates that the left hemisphere is in charge of abstract knowledge and expressing a particular meaning. Dall'Aglio (2019) also points out that we locate the symbolic in the prefrontal cortex (PFC) and temporo-parietal. "In the left temporo-parietal cortices the abstract signifying chain that gives structure to the subject is constituted. The PFC extends the symbolic through the internalized Other (parents) that establishes rules that organize discourse and actions" (p. 34). Through these cortices, processes of memory and active representation occur that, according to the author's Lacanian reading, account for the processes of significance. Damage in this area results in the emergence of contradiction.

Regarding the real register, Dall'Aglio (2019) proposes that the following should be agreed: the drive activates the detector of the need that derives in the homeostatic impulses originating in the upper brain stem. What is real is then the pressure of the drive in the internal organism that pushes toward the implementation of some type of action tending toward homeostasis. What Dall'Aglio does is divide this pressure of the drive into two: learned instincts, characterized by flexibility, without a universal consummatory goal, non-representational insistence, correspond to the real and are located in the diencephalon; and primitive, structural, object-specific instincts for the consummation of need are located in the brainstem. What is common between both types of instincts, says the author, is that each one "has its own homeostatic system of pleasure-displeasure" (p. 30). So the real is in the brain areas corresponding to affects, as opposed to the areas corresponding to representations, but it is associated with homeostasis and therefore depends on subcortical processes, being located in the "upper brainstem areas of bodily needs" (p. 34). In this sense, it corresponds to traces that are "before hippocampal memory (episodic, representational)" (p. 34). "Therefore, I suggest that the brainstem and the subcortical affective circuits are necessary structures that support the Lacanian real" (p. 30). Thus, being able to map the records in the brain, neuropsychoanalysis differs from psychoanalysis, which makes it impossible to achieve this from the subjective point of view: "the neuropsychoanalysis of affective instincts allows mapping the real with an impossible precision from the subjective point of view of psychoanalysis" (p. 35).

This contribution by Dall'Aglio establishes a Lacanian meta-neuropsychology that maps Johnston's location of the three registers in greater detail, through dialogue with the Freudian meta-neuropsychology of Mark Solms. It is not that Dall'Aglio has found the structural place in the brain of Lacan's registers, but that he has located them in the dynamics of their interrelationships. The real is present, but as negativity. It exists, but is missing from memory. Hence, he affirms that "the limit of reason is within reason" (Dall'Aglio, 2019, p. 28). However, at the clinical level, when a subject speaks, he does not say "this is the limit of what I know", but when he talks about what he feels he says "I don't know how to explain it to you". So instead of "the limit of reason is within reason" it is more fair to say "in reason there is a hole".

Psychoanalysis will disappear if it is not saved by neurobiology

The previous notions converge in a central thesis, of which they say that Freud would have been delighted: psychoanalysis will disappear if it is not saved by biology. Either psychoanalysis is integrated into the project of founding a biology of the mind or it will disappear. Kandel (2009a) affirms that psychoanalysis will remain blind if it does not adopt the point of view of biology and will disappear if it thinks of an unconscious isolated from the brain.

> Although psychoanalysis has historically been scientific in its aims, it has rarely been scientific in its methods. In fact, Freud and the original founders of psychoanalysis made few serious attempts to establish evidence for the efficacy of psychotherapy. That way of thinking changed in the 1970s, when Aaron Beck, a psychoanalyst at the University of Pennsylvania, set out to test Freud's insights on depression.
>
> (Kandel, 2018, p. 120)

For some reason Freud never made this attempt to prove the efficacy of psychoanalysis in terms of biology. They argue that Freud did not perform this test because he did not have the neuroimaging techniques available today. Insel (2009) affirms that "neurobiological tools are currently available to study the most unknown aspects of mental activity such as unconscious processes, emotions and impulses" (p. 32).

> In the first place, psychoanalysis has never had even the slightest scientific foundations. Furthermore, it also does not have a scientific tradition, using questioning to patients based not only on imaginative concepts but also on creative and critical experiments that were designed to explore, support, or as often happens, falsify these concepts. Many of the concepts of psychoanalysis are the product of clinical studies on isolated cases.
>
> (p. 58)

In truth, psychoanalysis does not have a scientific tradition because psychoanalysts question fixed, naturalized ideas. Psychoanalysis places the accent on a certain liberation in relation to the severity of the standards, the accent that Lacan placed on "his declared contempt for everything that was of the order of tradition" (Miller, 2015a, p. 151). While for these neurosciences tradition is essential. Indeed, they use the myth of Oedipus as a traditional organizer of the practice.

> It happens that brothers and sisters simply do not find the other an attractive sexual partner. This is an understatement: the mere thought of having sex makes them feel very uncomfortable or disgusted. (Children who grow up without siblings of the opposite sex do not understand emotion.) Freud argued that the existence of such a strong emotion is itself proof of the existence of an unconscious desire, especially when a man claims to be repulsed by the mere thought of having intercourse with his mother.
>
> (Pinker, 2001, pp. 584–585)

The use of suggestion is tradition in this method:

> [...] when you talk to someone and they listen, you not only establish visual and verbal contact, but the action of the neural mechanisms of the brain of the speaker have an evident and, supposedly, prolonged effect on the neural mechanisms of the listener, and vice versa.
>
> (Kandel, 2009b, p. 25)

That is, whoever listens receives the effects of whoever speaks. An alien I is incorporated into the I, as a result of which this first I behaves in certain respects like the other, assimilates it, and welcomes it within itself. "They are image therapies, which by this very fact are therapies for [...] identification with the master" (Miller, 1994b, p. 5). The psychoanalyst and his double. In the double there is a copy that reflects an image. That is the "copycat phenomenon [...] with which we now deal" (Miller, 2013b, p. 104). Freud absolutely discouraged this procedure:

> [...] I can assure you that you are misinformed if you suppose that advice and guidance in the affairs of life would be an integral part of analytical influence. On the contrary, we avoid such a mentoring role as far as possible; What we most desire is for the patient to make his decisions autonomously [...] Only in certain very young or totally defenseless and unstable people can we not respect this voluntary restriction. In them we are forced to combine the function of the doctor with that of the educator; but then we are fully aware of our responsibility, and we behave with the necessary caution.
>
> (Freud, 2011ñ, p. 394)

So, the notion of subject that Freud brings breaks the double relationship (Miller, 2015a). However, Talvitie (2009) denounces that "with the Freudian unconscious there is a notable and fundamental problem: no one has identified what kind of entity the unconscious is supposed to be" (p. 31).

> [...] psychoanalysis has effectively failed to take into account a large portion of the data that can be derived from our externally oriented perceptions. By limiting itself in this way to only half of the available data, psychoanalysis has not only restricted the possibilities of obtaining pure knowledge of the mental apparatus, but has relinquished our ability to apply this knowledge to influence the mental apparatus by physical means [...] Nothing will stop us from using this method to study the psychological functions that constitute the human mental apparatus as we conceive it in psychoanalysis and, therefore, from dynamically locating the main constituents of Freud's psychological model in the anatomical structures of the brain.
>
> (Solms and Solms, 2005, pp. 255–257)

For his part, Insel (2009) also states that psychoanalysis should import the model adopted in the natural sciences, otherwise it will not survive.

> The future of psychoanalysis, if it has one, passes through the context of empirical psychology, complemented by diagnostic imaging techniques, neuroanatomical methods, and human genetics. Within the framework of the sciences that study human cognition, psychoanalytic ideas can be scientifically verified, and this is where they can have the greatest impact.
>
> (p. 59)

Thus, the analyst, instead of entrusting himself to the suffering of the ethical subject and giving up being his master, will become the scientist guaranteed by neurobiology.

> Psychoanalysis should follow the direction opened by Lacan when he wondered about the existence of a science that included psychoanalysis. The neurosciences should find in psychoanalysis the necessary support points to orient themselves in the emergence of the unique, located within the general biological mechanisms discovered.
>
> (Ansermet and Magistretti, 2006, p. 25)

In this direction, Bazan and Zehetner (2018) affirm that the neurosciences have reached their golden age while psychoanalysis continues to struggle with its bad reputation. When Bazan is asked what led her down the path of biology, she replies that she read a lot of psychology books, saying "Well, the author may or may not be right, but I have no means of knowing" (p. 2). That is why we have majestic works (Bazan and Snodgrass, 2012; Bazan and Detandt, 2013; Bazan, 2011; 2012; Panksepp, 1998; Shevrin, 2003) that have strengthened their search for neuroscientific evidence of psychoanalytic concepts, and have worked to provide strong physiological frameworks for understanding Freudian concepts. The difference is that while some use the conceptual framework of psychoanalysis to interpret brain mechanisms, that is, they interpret neuroscience with psychoanalysis, such as Bazan and Dall'Aglio, others, on the contrary, extrapolate brain mechanisms to psychoanalytic theory, test psychoanalysis with neuroscience, placing neuro concepts as primordial, like Solms. But everyone is convinced that with neuroscientific findings, psychoanalysis will gain credibility.

According to the experiment by Bazan and Zehetner (2018), neuropsychoanalysts, in a careful procedure, select words that they consider important because they touch on a specific unconscious conflict for each participant in the experiment. They were able to verify that when the analysts were not present in the laboratory and presented these words subliminally to the participants within a millisecond, the participants had no idea of these words. They were able to record a wave, which is called an alpha brain wave, that was synchronized when the word about the unconscious conflict was presented. The alpha wave was only recorded when the patients were presented with the specific words. They came to the conclusion that, regardless of the analyst's presence or the transference relationship, an unconscious mental reality can be objectified, which is therefore not simply subjective (Bazan and Zehetner, 2018, p. 2).

From these studies, the biologization of the unconscious is equivalent to the independence of the analyst and the devaluation of the transference relationship. Bazan argues that one of Lacan's hypotheses is that the unconscious is structured like a language. And he says: "I want to test this hypothesis empirically" (p. 3). It is there when the unconscious will no longer be structured as a language, but will be the language, the metalanguage, and there we will lose our ethical position. The uselessness of the lack of organic substantiation of the unconscious is so disturbing for neuroscience that when Bazan is asked what question she would ask if she had the opportunity to speak with Sigmund Freud, she says that the topic would be "Sexuality, I would ask him, how come we have sexual fantasies that serve no purpose for evolutionary survival?" (p. 3).

Johnston (2013) claims that Freud's notion of the death drive is not clear, it is imprecise, it is scattered throughout Freud's later writings, it is messy and "it is actually a quasi-concept, an inconsistent jumble of phenomena that vaguely resemble each other (and occasionally even incompatible with each other)" (p. 61). He says that Freud was unable to present a clear and consistent metapsychological description of the notion of the death drive, and claims that this Freudian notion "names a set of unresolved problems rather than a polished and finalized conceptual solution" (p. 61). The good thing is that Ansermet and Magistretti (2010, 2012) were able to locate the drive between the brainstem and the insular cortex. Bazan and Detandt (2013) also claim that the concept of jouissance is "impermeable to understanding" (p. 1) and for this reason they qualify it as an "infamous" concept and one that should be dismissed if it is not saved by biology. Kandel (2009b) claims that psychoanalysis has not incorporated the methodologies and concepts of neurobiology "because it has not yet recognized itself as a branch of biology" (p. 71). In short, the list of neuroscientific authors who direct accusations toward Freud is endless.

The biologization of psychoanalysis was an iron law in 1950 when the condition to be a psychoanalyst was to have knowledge of neurobiology. But in 1890, very early in his work, Freud tells us that the curative value of the word resides in what he calls hopeful expectation, that is, trust in someone to whom the patient assumes power, an ability to heal. In the hopeful expectation, the person feels confidence in a force that comes from the one who carries out the treatment, an "effective force that strictly speaking we cannot fail to do without in all our attempts at treatment and cure" (Freud, 2011i, p. 121). Scientists and doctors themselves rely on this healing power of hopeful expectation. In this way, animic treatment is the oldest practice in the history of humanity, long before medical treatment. So, Freud concludes, scientific verification is not necessary to cure the pain of the soul:

> The confident expectation does not want to occur, but when the one who heals is not a doctor and can boast of ignoring everything regarding the scientific foundation of the art of healing, and when the resource used was not subjected to exact verification.
>
> (p. 122)

As if this were not enough, in 1926 Freud (2006b) said that lay people, non-medical people, can practice psychoanalysis. Another way of saying that psychoanalysis is not based on biological foundations.

Interview[1] with Dr. Carlos Daniel Mías[2]

Interview with Carlos Daniel Mías on Neurosciences and Psychoanalysis, conducted by Marco Máximo Balzarini on 4 January 2019.

MMB: In relation to the debate on psychoanalysis and neuroscience, I would like to know what you think about how neuroscience reads some topics. In particular, why are neurosciences more relevant, not only in groups of professionals who are starting out, but also within today's society?

CDM: There is an interesting metaphor to illustrate this debate. It was taught to me 30 years ago by the person with whom I was trained here in Córdoba, an eminence in this matter. It goes like this: In a dark street there was a lady who was turning in a circle around a lamppost. When a man passes he asks her: what are you doing? The lady replies "I'm looking for my son", the man asks her where she lost him? The lady points back into the darkness and answers "over there", to which the man asks "and why are you looking for him here?" And the lady answers "because this is where the light is". That is to say, psychoanalysis only searches where the theory gives it light, but ignores other things, which it does not take into account. In many forms of psychotherapy as well. What is not known is not explored or included in the analyses. Neuroscience, through the study of brain organization, has found numerous mental processes that have contributed to general psychology (for example, executive processes, decision-making, social cognition, or theory of mind, among others); others, on the other hand, have achieved greater empirical strength (for example, various forms of attention, memory, or emotional regulation). It can show results and is more practical, making it possible to identify processes that give more precision or efficiency to interventions. Furthermore, the purpose of science is to study and find out what different people have in common. There is no science of the individual. Psychoanalysis seems to make an excess of singularity, with a model somewhat closed to variables from other disciplines. It calls itself depth psychology, but it does not depart from its theoretical principles, it does not integrate with other disciplines. It must also be considered that many care practices are maintained to the extent that they generate work and money. For this reason, it is very difficult to review or update a theoretical model when it is part of a money income chain. It is necessary to know that neurosciences can address aspects that psychoanalysis does not know. We must recognize as psychoanalysts that the brain exists and sometimes it imposes conditions on you.

It is also necessary to be able to differentiate what is a theory from a doctrine. Theories are reviewed, updated, and integrated with other disciplines. The doctrines imply an adherence to emotional components, with knowledge that is not questioned. It is incredible that psychology is not taught to differentiate a theory from a doctrine, this is how many fall into indoctrination, where facts no longer matter, or where integration with other ideas seems to be heresy. On the other hand, neuroscience has an empirical position and contributes to the knowledge of mental processes that enrich any theory in psychology. Of course, it also lends itself to metaphors that move away from nervous representation and fall into generalities or falsehoods. For example, the brain can be programmed for success, or the left hemisphere is cold and the right is warm, or if the lip contracts to the right, the smile is false because it denotes activation of the left hemisphere, which is emotional. What does this mean? That metaphors, techniques, or resources that are far from empirical evidence frequently appear, that are imposed and expanded, that can generate demand in the mental health training market, but that demand then fades away. Those of us who are years old have seen many of these supposed theories go by, such as neurolinguistics, EMDR, brain reprogramming, and many interventions with the neuro suffix, which later end up falling apart. This seems to be the cycle of many theories that do not go beyond techniques or particular ways of interpreting, capable of selling courses everywhere, but which then disappear. Notice that biodecoding is now booming for many colleagues, which is just a metaphor and without scientific support. Or on other occasions, mindfulness appears as a technique that proposes to develop full awareness in the present and teaches meditation, something that may be fine for those who accept it and under certain conditions, but which is by no means a defined approach. Today they are in great demand, but surely in the coming years these approaches will no longer be talked about, and new ones will appear. Basically it seems that there is economic or market competition. Notice that psychologists are one of the few professions that live off of colleagues, through courses, or workshops or variants that always pay. This economic perspective of science, or rather, of health practices cannot be ignored. After all, much of science has been developed as a function of interests. And frequently politics also subsidizes with its interests or chains of favors.

MMB: In this sense of politics, when you read Manes, you also read a political position. Why are politics and neuroscience related?

CDM: Manes has achieved significant exposure, he is a great disseminator of neuroscience in our country, one of the most popular neurologists in Argentina. As a diffuser it is fine, thanks to this people are more interested today in the health of their brain and mind in a more integrated way. However, when disseminating to the general public, many metaphors are used and actions are justified based on them, but these often deviate from genuine brain functioning and may be functional to positions that are more ideological than scientific. This can lead to programs in education, with the intention of contributing precisely in a pragmatic sense. For example, neuroscientific applications have suggested that the practice of relaxation and meditation, as well as the recording of emotions and procedures for emotional regulation in schools, significantly reduce the problem of intra-school violence. However, a leader or school director prefers a program that is clear, simple, and operational, and where possible with immediate results. Not one that questions or reflects on concepts that can hardly be understood, and is useless for interventions that refer to groups of subjects. It is known that psychologists deal with things that are difficult to operationalize, with abstractions, hence the low efficiency of many interventions in psychotherapy. It's like the problem of dealing with self-esteem. This is just an abstract concept, and when people want to put it into practice they don't know how to do it. So the neurosciences provide processes that make it possible to better identify the construction of self-esteem; for example, training behavioral inhibition resources, planning and execution, decision-making, social cognition, or emotional regulation; that are finally the source of an emotional experience of achievement, self-control and self-esteem. We cannot say exactly how the brain operates in each case, but using research we can give an approximation. Neuroscience bases its pragmatic position by saying that with such treatment students will be able to achieve assertive behaviors. The problem is that it can fall into reductionist positions, with the labeling of children for example, and fall into a theoretical spiral that moves away from empirical bases. In science, as in the politics with which it is linked, there are also emotions, prejudices, interests, and chains of favors.

MMB: Lacan criticizes science that forecloses the subject, that is, it does not take it into account. What does neuropsychology say about this?

CDM: Before, everyone was with the unconscious, and it was a way of penetrating the source of their problems into the social imaginary. Neurosciences are changing that representation, and today people are increasingly interested in knowing how the brain works and how it orients itself in life. It is being understood that knowing how the brain works is also knowing how the mind operates, and how we can know ourselves. Neuropsychology, for its part, also deals with how the brain learns and works, but in a non-individual scientific way. There is no science of the individual. Anthropologist Clyde Kluckhohn says that people are in some ways like everyone else, in some ways like some, and in some ways not like any body. Science points to what we have in common with some people (for example, a trait, disease, belief, or behavior). Thus, science has to provide answers to cure schizophrenia as a disease, but not Juan Perez's schizophrenia. I repeat, there is no science of the individual. Each professional has to know how to adapt to the singular. When you run intervention programs, you are supposed to empirically show that it favors an important part of society. For this reason, public services tend to be protocolized, assuming that not everyone can benefit. But in private services, the model can be centered on the patient, with longer times and adjustments, with greater tolerance and consideration of multiple factors, but always with procedures validated in some way. For this reason, professionals know that we are responsible for the procedures used, but not for the results. Neuropsychology adapts to these two aspects, since it is accustomed to interdisciplinarity, as was recognized just a few years ago by the Argentine Mental Health Law. Basic neuroscience and cognitive neuroscience are sciences that try to understand how the brain works, but do not make intervention applications, but rather experimental ones. They can provide some empirical foundations. Neuropsychology is an applied discipline of neurosciences. It makes multifactorial models of health and disease prevail. The genetic component, heredity, neurochemical dysfunction, and neurocognitive functions according to a model of brain architecture are considered, but the psychological part is also seen, how it is structured, its family context, its functional context, its evolutionary stage of life, and environmental stimulating factors. An analysis as integrated as possible is attempted. It can be said that a family has the anxiety gene, and that the members will have it, but we are not left with that. Because we can't do something with it. We also know that there are epigenetic factors. The power of multifactoriality is seen in the ability to explore different causes of behavior. For example, a patient who does not speak fluently or has a low register of their emotional or cognitive states could be due to many factors, but without doing a genetic study we can also consider that they have frontal hypoflow, that is, that their cerebral metabolism in critical areas of the brain works in a diminished or sluggish manner. Later, more objective studies will be necessary to support this hypothesis, such as neuroimaging. Psychoanalysis, like psychology in general, explains what the theory sheds light on, and it is very likely that it would be thinking about the unconscious. This is the case of the so-called "absent subjects" in therapy, who are often explained by defensive or unconscious mechanisms, but objective studies can show that they have hypoflow in the brain, among many possibilities. Is that unconscious? We agree that if problems are rooted out it is better, but then it must be discussed and objectively demonstrated where the root of a problem is. For psychoanalysis, it seems that the root is always in the same place, or at least it is of the same nature. Is this unique? However, sometimes it may be the unconscious, other times it may be in the limitations of the brain, or epigenetic factors. For example, a child who does not work or learn at school, you can see his school context, his family context, and his neurocognitive processing, among many others. And then we see which factor is of more weight, to be addressed clinically. On the other hand, sometimes you have to be strategic in your intervention, for example, start intervening more from the periphery and not so directly on the problem. And this does not exclude that other variables do not continue to be treated. But psychoanalysis, like other orientations in psychology, seems to have difficulty integrating variables outside its own discipline. They speak of what they know, illuminate only what they know, and do not speak of what they do not know. Psychoanalysis sheds light on only one aspect, but neglects other aspects, where it cannot see. In practice, it appears to be more of a doctrine than a theory. In neuropsychology we teach that when you have a hammer as your only tool (theory) you tend to see all problems as the head of a nail. And this goes against the grain even with the mental health law.

MMB: There is a current debate around the power of the pharmaceutical industries and the medicalization of children. How do you understand it?

CDM: First of all, you cannot be for or against medicines. This would be an ideological position. Rather, you have to be professional, and determine when one intervention or another is justified. For example, if you have a boy who is impulsive, aggressive, violent, lacking self awareness because he places himself in risky situations, and who meets the criteria for true ADHD (which includes frontal hypoflow and executive deficits), the application of a drug may be beneficial for that patient. The drug contributes to the treatment, it must be based on therapeutic objectives. Rarely drugs can cure by themselves, so pharmacology is integrated into a work program. They cannot be parallel actions. The clinical psychologist or some psychoanalysts sometimes notice the need for a drug, but they work in parallel, as if drugs did not affect mental processes, they follow their own objectives. Laboratories are significantly improving the quality of life of patients, not because they cure them, but because they affect certain cognitive processes, which includes that they benefit much more from psychotherapy. And we psychologists have to worry about that. I remind you that in the 1960s it was thanks to the appearance of chlorpromazine that bloody treatments were abandoned, outpatient treatments became possible, and numerous psychiatric clinics closed for this reason. Today you don't have to be sick to take a medicine, but you act preventively, taking care of the brain, protecting it from intense emotions, the toxic effects of depression, bad habits, or stress. Vitamins, dietary complexes, vaccinations. If the patient comes with addiction, we see all the factors, in multiple ways, that intervene in its development. Addictions affect neurons and produce deficits. This must be taken into account. Note that a psychologist position assumes that a person consumes illegal drugs to evade reality, but if this is so, then comes the problem of which reality they evade? There are many young people with a history of cognitive and learning deficits, who for these reasons end up being bullied or having low self-esteem, with feelings of uselessness, which drugs later allow them to forget. But sometimes the problem has been primarily nervous, whether due to maturational delays, nutritional deficiencies, stimulation, or simply perinatal problems, among many possibilities. Then, it must be considered that regardless of how consumption was reached, once there is chronicity with drugs, you already have an affectation on the nervous system and cognitive processes, which will impact behavior. Now you have two problems and not just one. And both deserve to be treated equally.

MMB: The *Diagnostics and Statistics Manual* has recently been published in its fifth revision. What do you think of this permanent update of diagnoses and mental nosological entities?

CDM: The DSM is a coder or nomenclator of mental illnesses, and it does so in a descriptive way in order to establish consensus worldwide, and promote joint research so that all researchers in the world know what they are talking about. Of course they need to be updated. They are updated as new clinical entities exist. Of which we have to have a criterion. For example, now gender dysphoria or attachment problems have appeared and we have to know what that means. I insist that the DSM is only a nomenclator, which aims to unify criteria for research; like the ICD-10. Even social works demand its use. The problem is that many psychologists or psychiatrists have taken it as a psychopathology manual, and nothing further from its purpose. But this is a training problem, not the DSM. The problem that I see is that they encourage categorical readings, when in reality we see the need for dimensional perspectives in psychotherapy. In addition, they do not include possible transdiagnostic perspectives, based on common denominators that have various diseases or psychological states. Perhaps for this reason the APA (American Psychological Association) has distanced itself from the DSM5, although I do not rule out that it is also due to economic interests and the I market. As in politics, we will never know the true motivations of decisions or laws, we just have to settle for their necessity and actuality.

MMB: I understand that neuropsychology uses Freud to make interpretations about psychic phenomena. Specifically, how would you say that neuropsychology uses Freud?

CDM: If we neuropsychologists were psychoanalysts, we would surely be Freudians. We have more affinity with Freud. He was a doctor, and at a time when there were no tools to study the integrated organism and mind, he had to invent one: the Freudian psychic apparatus. Even with topics that remind us of the phylogenetic developments of the nervous system. Then came the divergences in psychoanalysis, many times completing Freud's theory, other times in defense of personal egos.

Those were times when many wanted to be parents of a theory. Then came Lacan. We do not understand from neuropsychology how Lacan tells his students that he is Freudian, instead of Lacanian. It's confusing. We cannot follow Lacan. In my opinion, it is an abstraction, closer to being a doctrine than a theory. For this reason, I always recommend the book *El olvido de la razón* [The forgetfulness of reason] by Sebrelli, an Argentine philosopher, who shows how theories developed by great thinkers of the 20th century end up becoming doctrines.

MMB: In relation to autism, in psychoanalysis there is doubt about its etiology and research continues. How do you understand it?

CDM: Autism today has more than 40 anomalies in the nervous system found and proven. Advanced neuroimaging and laboratory techniques have been demonstrating this, although a single pattern of involvement has not yet been recognized. It is estimated that anomalies will continue to be found. We cannot ignore the biological contribution to the etiology of autism. The tendency is to look for the biological bases, in order to understand more of the mental processes. If psychoanalysis had a better explanation with consistent results, it would be overwhelmingly reflected in scientific databases. It is like the problem of Alzheimer's, where I have also read non-biological interpretive psychoanalytic explanations. Well, this is where the discussions end, they don't make sense. What we do depends on what our theory tells us about how things are; what happens next depends on how things are, not how we think they are. For this reason, in science it is always interesting what can be seen as a result of the theories, and that naturally the experiences can be replicated.

MMB: Thank you very much Daniel!

Notes

1   Reviewed and authorized by the interviewed to be published.
2   Doctor in Health Sciences; Director of the Neuropsychology Service, UNC Faculty of Psychology; Full Professor Neuropsychology Chair, UNC; Adjunct Professor Chair Neurophysiology and Psychophysiology, UNC; Professor and Member of the Academic Committee of the Master's Degree in Clinical Psychology, Integrative Cognitive mention, UNSL (Coneau Accredited); Professor and Member of the Academic Committee of the Master's Degree in Neuropsychology, UFLO (Coneau Accredited), Director of the Cognitive and Behavioral Psychotherapy Postgraduate Program. Integrated Methods and Applications. Faculty of Psychology, UNC.

References

Aguiar, A. (2018). The "real without law" in psychoanalysis and neurosciences. *Frontiers in Psychology*, 9: 851. doi: 10.3389/fpsyg.2018.00851
Ansermet, F. and Magistretti, P. (2002). Introducción. In Nathalie Georges, Nathalie Marchaison, and Jacques-Alain Miller, *De las neurociencias a las logociencias, ¿Quiénes son sus psicoanalistas?* Paris: du Seuil.

Ansermet, F. and Magistretti, P. (2006). *A cada cual su cerebro. Plasticidad neuronal e inconsciente.* Buenos Aires: Katz.

Ansermet, F. and Magistretti, P. (2010). L' "île" de la pulsion. In *Les énigmes du plaisir.* Paris: Odile Jacob.

Ansermet, F. and Magistretti, P. (2012). The island of the drive. *Swiss Archives of Neurology and Psychiatry,* 163 (8), 281–285.

Ayan, S. (2006). Mecanismos del inconsciente. In *Mente y cerebro. Freud. Investigación y ciencia* (18), 62–67.

Balzarini, M. (2021). La formación en psicoanálisis de orientación lacaniana y en neurociencias psicoanalíticas. In *Escritos de Posgrado,* año 1, Nº 3: Facultad de Psicología, Universidad Nacional de Rosario. Accessed 23 September 2021 from: https:// escritosdeposgrado-fpsico.unr.edu.ar/?p=377

Bandler, R. and Grinder, J. (1980). *La estructura de la magia. Lenguaje y terapia.* Santiago de Chile: Cuatrovientos.

Barondes, S. (2004). Nuevas píldoras para la mente. In *Revista Lacaniana. Las prácticas de la escucha y sus argumentos* (2). Buenos Aires: EOL.

Barros, M. (2004). La salud de los nominalistas. Un estudio sobre las prácticas psicoterapéuticas. In *Revista Lacaniana. Las prácticas de la escucha y sus argumentos* (2). Buenos Aires: EOL.

Bassols, M. (2011a). *Tu yo no es tuyo.* Buenos Aires: Tres Haches.

Bassols, M. (2011b). Las neurociencias y el sujeto del inconsciente. Conferencia pronunciada en Granada. Instituto del Campo Freudiano. Accessed from: www.icf-granada. net/2012-04-04-08-33-03/videos/83-las-neurociencias-y-el-sujeto-del-inconsciente

Bassols, M. (2012b). Psicoanálisis, sujeto y neurociencias. Presentación del libro *Sutilezas analíticas* en Alianza Francesa de San Ángel. Nueva Escuela Lacaniana. Mexico D.F.

Bassols, M. (2013a). La vigencia del psicoanálisis. Interview by Bordon, J.M. in Magazine *Noticias.* (pp. 118–120). Centro de investigación y docencia en psicoanálisis. Lima. Accessed from: www.enapol.com/images/Prensa/13-12-06_Entrevista-a-Miquel-Bassols.pdf

Bassols, M. (2013b). La diferencia entre psicoanálisis y ciencia en torno a la idea de cuerpo. Entrevista en *EOL Rosario* por Manuel Ramírez. Accessed from: www.eol. org.ar/template.asp?Sec=prensa&SubSec=america&File=america/2013/13-12-01_ Entrevista-a-Miquel-Bassols.html

Bassols, M. (2014). El ocaso de la psiquiatría, ¿y después? In *Freudiana* (72). ELP de la EFP miembro de la AMP. Catalunya: Repro Disseny.

Bassols, M. (2016b). La fascinación mecánica. In *Freudiana* (77–78). ELP de la EFP miembro de la AMP. Catalunya: Repro Disseny.

Bassols, M., Laurent, E., and Berenguer, E. (2006). Lost in cognition. In *Freudiana* (46). ELP de la EFP miembro de la AMP. Catalunya: Repro Disseny.

Bazan, A. (2006). Primary process language. *Neuro-Psychoanalysis,* 8, 157–159.

Bazan, A. (2011). Phantoms in the voice: a neuropsychoanalytic hypothesis on the structure of the unconscious. *Neuropsychoanalysis,* 13 (2), 161–176.

Bazan, A. (2012). From sensorimotor inhibition to Freudian repression: insights from psychosis applied to neurosis. *Front. Psychol.,* 3, 452.

Bazan, A. and Detandt, S. (2013). On the physiology of jouissance: interpreting the mesolimbic dopaminergic reward functions from a psychoanalytic perspective. *Frontiers in Human Neuroscience,* 7 (709). doi:10.3389/fnhum.2013.00709.

Bazan, A., Detandt, S., and Van de Vijver, G. (2017). The mark, the thing, and the object: On What commands repetition in Freud and Lacan. *Frontiers in Psychology,* 8 (22). doi: 10.3389/fpsyg.2017.02244.

Bazan, A., Shevrin, H., Brakel, L., and Snodgrass, M. (2007). Motivations and emotions contribute to a-rational unconscious dynamics: evidence and conceptual clarification. *Cortex*, 43 (8), 1104.

Bazan, A. and Snodgrass, M. (2012). On unconscious inhibition: Instantiating repression in the brain. In A. Fotopoulou, D.W. Pfaff, and E.M. Conway (eds.), *Trends in Psychodynamic Neuroscience*. Oxford: Oxford University Press, pp. 307–337.

Bazan, A., Van Draege, K., De Kock, L., Brakel, L., Geerardyn, F., and Shevrin, H. (2011). Empirical evidence for Freud's theory of primary process mentation in acute psychosis. *Psychoanalytic Psychology*. Advance online publication. Doi: 10.1037/a0027139

Bazan, A. and Zehetner, S. (2018). When people recount their dreams, they don't talk about the hippocampus. In *The Vienna Psychoanalyst*. Accessed 17 April 2023 from: www.theviennapsychoanalyst.at/index.php?start=2&wbkat=8&wbid=1112

Beck, A., Rush, J. Shaw, B., and Emery, G. (1983). *Terapia cognitiva de la depresión*. Bilbao: Desclee de Brouwer.

Benjamin, W. (1982). *Experiencia y pobreza*. Trad. Jesús Aguirre. Madrid: Taurus.

Binder, J. and Desai, R. (2011). The neurobiology of semantic memory. *Trends in Cognitive Sciences*, 15 (11), 527–536.

Born, J. and Wagner, U. (2006). ¿Sueñan las redes neuronales?. In *Mente y cerebro. Freud. Investigación y ciencia* (18), 68.

Brakel, L. and Shevrin, H. (2005). Anxiety, attributional thinking, and the primary process. *International Journal of Psycho-Analysis*, 86 (6), 1679–1693.

Buchheim, A., Cierpka, M., Kächele, H., and Roth, G. (2013). Efectos del psicoanálisis en el cerebro. In *Mente y cerebro. El legado de Freud*. La neurociencia demuestra la eficacia del psicoanálisis. *Investigación y ciencia* (62), 26–29.

Castanet, H. (2023). *Neurología versus psicoanálisis*. Buenos Aires: Grama Navarin.

Chamorro, J. (2011). *¡Interpretar!* Buenos Aires: Grama.

Copjec, J. (2015). *Read My Desire: Lacan Against the Historicists*. 2nd ed. London: Verso Press.

Cuñat, C. (2019). El paradigma neurocientífico y el imaginario social. In *Freudiana* (86) "Inconsciente y cerebro: nada en común". ELP de la EFP miembro de la AMP. Catalunya: Repro Disseny.

Dall'Aglio, J. (2019). Of brains and Borromean knots: A Lacanian meta-neuropsychology. *Neuropsychoanalysis*, 21 (1), 23–38, doi: 10.1080/15294145.2019.1619091

Dall'Aglio, J. (2020a). No-Thing in common between the unconscious and the brain: on the (im)possibility of Lacanian Neuropsychoanalysys. *ResearchGate, Psychoanalysis Lacan*, 4. Accessed 15 April 2023 from: https://researchgate.net/publication/342870600

Dall'Aglio, J. (2020b). Sex and prediction error, part 2: Jouissance and the free energy principle in neuropsychoanalysis. *Japa*, 69 (4), 715–741. doi: 10.1177/00030651211042377

Dall'Aglio, J. (2021). What can psychoanalysis learn from neuroscience? A theoretical basis for the emergence of a neuropsychoanalytic model. *Contemporary Psychoanalysis*, 57 (1), 125–145, doi: 10.1080/00107530.2021.1894542

Damásio, A. (1994). *El error de Descartes. La razón de las emociones* [*Descartes' Error*]. Buenos Aires: Andres Bello.

Davidovich, M. y Winograd, M. (2010). Psicoanálisis y neurociencias: un mapa de los debates. *Psicologia em Estudo*, 15 (4), 801–809.

De Georges, P. (2005). Paradigma de desencadenamiento. En Jacques-Alain Miller y otros, Los inclasificables de la clínica psicoanalítica, Buenos Aires, *Paidós*, p. 41–46.

Decety, J. (1996). Neural representations for action. *Rev. Neuroscience*, 7, 285–297

Dehaene, S. (2015). *La conciencia en el cerebro. Descifrando el enigma de cómo el cerebro elabora nuestros pensamientos*. Buenos Aires: Sigloveintiuno.

Delgado, S., Strawn, J., and Pedapati, E. (2015). *Contemporary Psychodynamic Psychotherapy for Children and Adolescents. Integrating Intersubjectivity and Neuroscience*. Berlin: Springer.

Deneke, F.-W. (2006). Un modelo estructural revisado. In *Mente y cerebro. Freud. Investigación y ciencia* (18), 71.

Edelman, G. and Tononi, G. (2002). *El universo de la conciencia. Cómo la materia se convierte en imaginación*. Barcelona: Crítica.

Eichenbaum, H., Cahill, L., Gluck, M., Hasselmo, M., Keil, F., Martin, A., and Williams, C. (1999). Learning and memory: systems analysis. In Zigmond, M., Bloom, F., Landis, S., Roberts, J. and Squire, L (eds.), *Fundamental Neuroscience*. New York: Academic Press, pp. 1455–1486.

Ellis, A. (2000). *Vivir en una sociedad irracional: Una guía para el bienestar mediante la terapia racional-emotivo-conductual*. Barcelona: Paidós.

Fajnwaks, F. (2019). De la industria de la imagen a los usos del soñar. In *Freudiana* (86) Inconsciente y cerebro. Escuela Lacaniana de Psicoanálisis de Catalunya.

Freud, S. [1926] (2006b). *¿Pueden los legos ejercer el análisis?* In *Sigmund Freud. Obras Completas*. Tomo XX. Buenos Aires: Amorrortu.

Freud, S. [1888] (2011a). Histeria. In *Obras Completas*. Tomo I. Buenos Aires: Amorrortu.

Freud, S. [1930] (2011b). El malestar en la cultura. In *Sigmund Freud. Obras Completas*. Tomo XXI. Buenos Aires: Amorrortu.

Freud, S. [1905] (2011d). Tres ensayos de teoría sexual. In *Sigmund Freud. Obras Completas*. Tomo VII. Buenos Aires: Amorrortu.

Freud, S. [1896] (2011g). Carta 52. In *Sigmund Freud. Obras Completas*. Tomo I. Buenos Aires: Amorrortu.

Freud, S. [1890] (2011i). Tratamiento psíquico (tratamiento del alma). In *Sigmund Freud. Obras Completas*. Tomo I. Buenos Aires: Amorrortu.

Freud, S. [1893] (2011j). Algunas consideraciones con miras a un estudio comparativo de las parálisis motrices orgánicas e histéricas. In *Sigmund Freud. Obras Completas*. Tomo I. Buenos Aires: Amorrortu.

Freud, S. [1916] (2011l). 18ª conferencia. La fijación al trauma, lo inconciente. In *Sigmund Freud. Obras Completas*. Tomo XVI. Buenos Aires: Amorrortu.

Freud, S. [1905] (2011ll). Fragmento de análisis de un caso de histeria. In *Sigmund Freud. Obras Completas*. Tomo VII. Buenos Aires: Amorrortu.

Freud, S. [1917] (2011ñ). 27ª conferencia. La trasferencia. In *Sigmund Freud. Obras Completas*. Tomo XVI. Buenos Aires: Amorrortu.

Freud, S. [1892] (2011o). Prólogo y notas de la traducción de J.-M. Charcot, *Leçons du mardi de la Salpêtrière*. In *Sigmund Freud. Obras Completas*. Tomo I. Buenos Aires, Argentina: Amorrortu.

Freud, S. [1915] (2012b). Lo inconsciente. In *Sigmund Freud. Obras Completas*. Tomo XIV Buenos Aires: Amorrortu.

Freud, S. [1900] (2012c). La interpretación de los sueños. Cap. VII. In *Sigmund Freud. Obras Completas*. Tomo V. Buenos Aires: Amorrortu.

Freud, S. [1915] (2012g). Pulsiones y destinos de pulsión. In *Sigmund Freud. Obras Completas*. Tomo XIV Buenos Aires: Amorrortu.

Freud, S. [1920] (2012h). Más allá del principio de placer. In *Sigmund Freud. Obras Completas*. Tomo XVIII. Buenos Aires: Amorrortu.

Freud, S. [1913] (2012k). Sobre la iniciación del tratamiento. In *Obras Completas*. Tomo XII. Buenos Aires: Amorrortu.

Freud, S. [1914] (2012ll). Introducción del narcisismo. In *Sigmund Freud. Obras Completas*. Tomo XIV Buenos Aires: Amorrortu.

Fridman, P. and Millas, D. (2005a). La exaltación maníaca. Las muertes del sujeto. In Jacques-Alain Miller et al., *Los inclasificables de la clínica psicoanalítica*. Buenos Aires: Paidós, pp. 81–87.

Fridman, P. and Millas, D. (2005b). Segunda discusión. La muerte del sujeto. In Jacques-Alain Miller et al., *Los inclasificables de la clínica psicoanalítica*. Buenos Aires : Paidós, pp. 89–98.

Gallese, V. (2000). The inner sense of action: agency and motor representations. *J. Conscious. Stud.*, 7, 23–40.

Grinbaum, G. (2022). El hijo adolescente de Harry Potter. In *Rayuela* (9). Accessed 23 November 2022 from: www.revistarayuela.com/es/009/template.php?file=notas/de-padres-e-hijos-en-el-mundo-de-la-inexistencia-del-otro.html

Han, B.-C. (2012). *La sociedad del cansancio*. Barcelona: Herder.

Han, B.-C. (2022). *Capitalismo y pulsión de muerte*. Barcelona: Herder.

Ibáñez, A. (2017). ¿Qué son las neurociencias? Noche de la EOL. In *e-Mariposa* (10). Temas de psiquiatría y psicoanálisis. Revista del Departamento de Estudios sobre Psiquiatría y Psicoanálisis (ICF-CICBA). Buenos Aires: Grama, pp. 25–31

Insel, T. (2009). Un nuevo marco intelectual para la psiquiatría. In E. Kandel (ed.), *Psiquiatría, psicoanálisis, y la nueva biología de la mente*. Tercera ed. España, Barcelona: Ars Medica.

Johnston, A. (2013). Drive between brain and subject: an immanent critique of Lacanian neuropsychoanalysis. *The Southern Journal of Philosophy*, 51 (Spindel Supplement), 48–84. doi: 10.1111/sjp.12019

Jones, E. (1981). *Vida y obra de Sigmund Freud*. Tomo 1. Barcelona: Anagrama.

Judith, L. and Rapoport, M. (2009). La psicoterapia y la sinapsis única. In E. Kandel (ed.), *Psiquiatría, psicoanálisis, y la nueva biología de la mente*. Tercera ed. España, Barcelona: Ars Medica.

Kafka, F. (2011). *Informe para una Academia*. Buenos Aires: Maldoror.

Kandel, E. (2001a). Psychotherapy and the single synapse: the impact of psychiatric thought on neurobiological research. *Neuropsychiatry Clin, Neuroscience*, 13 (2), 290–300.

Kandel E. (2001b). The molecular biology of memory storage: a dialogue between genes and synapses. *Science*, 294, 1030–1038.

Kandel, E. (2007). *En busca de la memoria. El nacimiento de una nueva ciencia de la mente*. Buenos Aires: Katz conocimiento.

Kandel, E. (2009a). Aspiraciones de la biología para un nuevo humanismo. In E. Kandel (ed.), *Psiquiatría, psicoanálisis, y la nueva biología de la mente*. Tercera ed. España, Barcelona: Ars Medica.

Kandel, E. (2009b). La influencia del pensamiento psiquiátrico en la investigación neurobiológica. In E. Kandel (ed.), *Psiquiatría, psicoanálisis, y la nueva biología de la mente*. Tercera ed. España, Barcelona: Ars Medica.

Kandel, E. (2018). *The Disordered Mind*. Nueva York: Farrah, Strauss and Giroux.

Kandel, E., Schwartz, J., and Jessell, T. (2001) *Principios de neurociencia*. Cuarta ed. España: McGraw Hill Interamericana España.

Kernberg, O. (1998). *Love Relations*. Inglaterra: Yale University Press.

Lacan, J. [1975–1976] (2006a). *El Seminario. Libro 23. El sinthome*. Buenos Aires: Paidós.

Lacan, J. [1969–1970] (2007). *El Seminario. Libro 10. La angustia*. Buenos Aires: Paidós.

Lacan, J. [1972–1973] (2008f). *El Seminario. Libro 20. Aun*. Buenos Aires: Paidós.

Lacan, J. [1955–1956] (2009f). *El Seminario. Libro 3. Las psicosis*. Buenos Aires: Paidós.

Lacan, J. [1953–1954] (2012b). *El seminario. Libro 1. Los escritos técnicos de Freud*. Buenos Aires: Paidós.

Lacan, J. [1973] (2012f). Televisión. In *Otros escritos*. Buenos Aires: Paidós.

Langaney, A. (2006). El sentido de la seducción. In *Mente y cerebro. Freud. Investigación y ciencia* (18), 80–82.

Lardjane, R. (2019). El inconsciente y el cerebro en psiquiatría. In *Lacan cotidiano. Para Pipol 9*. Revista de Psicoanálisis (824). BOLC.

Laurent, E. (2002). *Síntoma y nominación*. Buenos Aires: Diva.

Laurent, E. (2005). *Lost in cognition. El lugar de la pérdida en la cognición*. Buenos Aires: Diva.

Laurent, E. (2006). "Principios rectores del acto analítico". In *Mediodicho N° 31*. Córdoba: EOL Sección Córdoba.

Laurent, E. (2007). ¡Es difícil no estar deprimido! Reportaje por Magdalena Ruiz Guiñazu. In EOL Publicaciones. Accessed 11 April 2019 from: http://www.eol.org.ar/template.as p?Sec=prensa&SubSec=america&File=america/2007/07_12_09_laurent_reportaje.html

Laurent, E. (2008). Usos de las neurociencias para el psicoanálisis. Comunicación en el coloquio organizado en el Collège de France por Pierre Magistretti, con el título "Neurociencias y psicoanálisis". Accessed 23 August 2019 from: www.wapol.org/es/articulos/ Template.asp?intTipoPagina=4&intPublicacion=5&intEdicion=43&intIdiomaPublicacio n=1&intArticulo=1447&intIdiomaArticulo=1

Laurent, E. (2010). Interpretar la psicosis. In Cuadernos del Instituto Clínico de Buenos Aires. No. 13.

Laurent, E. (2011). La ilusión del cientificismo, la angustia de los sabios. In *Freudiana* (62). ELP de la EFP miembro de la AMP. Catalunya: Repro Disseny.

Laurent, E. (2020b). *El nombre y la causa. Conicet y UNC*. Córdoba: IIPsi Instituto de Investigaciones Psicológicas

McGilchrist, I. (2009). *The Master and his Emissary: The Divided Brain and the Making of the Western World*. New Haven: Yale University Press.

Mías, C. (2008). *Principios de neuropsicología clínica con orientación ecológica. Aspectos teóricos y procedimentales*. Córdoba: Encuentro.

Miller, J.-A. (1994b). Psicoterapia y psicoanálisis. In *Revista Freudiana* (10). Escuela Europea de Psicoanálisis-Catalunya.

Miller, J.-A. [1995] (1996c). El olvido de la interpretación. In *Entonces: Shhh [...]* Buenos Aires: Minilibros Eolia.

Miller, J.-A. (2004b). Improvisación sobre *Rerum Novarum*. In *Revista Lacaniana. Las prácticas de la escucha y sus argumentos* (2). Buenos Aires: EOL.

Miller, J.-A. (2004c). Verdad, probabilidad estadística, lo real. In *Revista Lacaniana. Las prácticas de la escucha y sus argumentos* (2). Buenos Aires: EOL.

Miller, J.-A. [1987] (2006). *Introducción al método psicoanalítico*. Buenos Aires: Eolia-Paidós.

Miller, J.-A. [1985–1986] (2010a). *Extimidad*. Buenos Aires: Paidós.

Miller, J.-A (2010b). *Efecto retorno sobre la psicosis ordinaria*. En Freudiana, Revista Psicoanalítica publicada en Barcelona bajo los auspicios de la Escuela Lacaniana de Psicoanálisis, N°58, enero/abril 2010.

Miller, J.-A. [1998–1999] (2011a). Paradigmas del goce. In *La experiencia de lo real en la cura psicoanalítica*. Buenos Aires: Paidós.

Miller, J.-A. [2000–2001] (2013b). *El lugar y el lazo*. Buenos Aires: Paidós.

Miller, J.-A. [2006–2007] (2014b). *El ultimísimo Lacan*. Buenos Aires: Paidós.

Miller, J.-A. [2008] (2015a). *Todo el mundo es loco*. Buenos Aires: Paidós.

Miller, J.-A. [1984–1985] (2021d). *1, 2, 3, 4*. Buenos Aires: Paidós.

Panksepp, J. (1998). *Affective Neuroscience: The Foundations of Human and Animal Emotions*. New York: Oxford University Press.

Pina, A. (2008). *Psiquiatría y psicoanálisis en el marco de las neurociencias*. Barcelona: Biblioteca nueva.

Pinker, S. (2001). *Cómo funciona la mente*. Barcelona: Destino.

Pinker, S. (2004). Órganos de computación. Entrevista realizada por Brockman. In *Revista Lacaniana. Las prácticas de la escucha y sus argumentos* (2). Buenos Aires: EOL.

Price, C., Moore, C., Humphreys, G., and Wise, R. (1997). Segregating semantic from phonological processes during reading. *Journal of Cognitive Neuroscience*, 9 (6), 727–733.

Provine, R. (2006). El Bostezo. In *Mente y cerebro. Freud. Investigación y ciencia* (18), 17–25.

Rapcsak, S., Beeson, P., Henry, M., Leyden, A., Kim, E., Rising, K., and Cho, H. (2009). Phonological dyslexia and dysgraphia: cognitive mechanisms and neural substrates. *Cortex*, 45 (5), 575–591.

Redmond, J. (2015). Debating the subject: Is there a Lacanian neuropsychoanalysis? *Psychoanalysis Lacan*, 1. Accessed 9 May 2023 from: https://lacancircle.com.au/wp-content/uploads/2020/09/Debating_the_subject.pdf

Rosales, J. (2017). *La valía de la escritura testimonial para la enseñanza psicoanalítica*. Querétaro, México: Fontamara.

Shevrin, H. (2003). The psychoanalytic theory of drive in the light of recent neuroscience findings and theories. 1st Annual C. Philip Wilson M.D. Memorial Lecture, New York.

Simonet, P. (2019). Claridad hipnótica del cerebro. In *Lacan cotidiano. Para Pipol 9*. Revista de Psicoanálisis (824). BOLC.

Slezak, D. (2018). Una App que ayuda a diagnosticar esquizofrenia a través del análisis del discurso de pacientes. Conicet. UBA. Argentina. Accessed from: www.conicet.gov.ar/una-app-que-ayuda-a-diagnosticar-esquizofrenia-a-traves-del-analisis-del-discurso-de-pacientes/

Solms, K. and Solms, M. (2005). *Estudios clínicos en neuropsicoanálisis. Introducción a la neuropsicología profunda*. Bogotá: Fondo de cultura económica.

Solms, M. (2004). Psychanalyse et neurosciences. *Pour la science*, 324.

Solms, M. (2006). Neuropsicoanálisis. Entrevista por Steve Ayan. In *Mente y cerebro. Freud. Investigación y ciencia* (18), 74.

Solms, M. (2007). Sigmund Freud hoy. *Revista Psicoanálisis*, 5, 115–119.

Solms, M. (2013). The conscious id. *Neuropsychoanalysis*, 15 (1), 5–19. Accessed 16 April 2023 from: https:// doi.org/10.1080/15294145.2013.10773711

Solms, M. (2015). *The Feeling Brain: Selected Papers on Neuropsychoanalysis*. London: Karnac.

Solms, M. (2017a). What is "the unconscious", and where is it located in the brain? A neuropsychoanalytic perspective. *Annals of the New York Academy of Sciences*, 1406(1), 90–97.

Solms, M. (2017b). "The unconscious" in psychoanalysis and neuroscience: An integrated approach to the cognitive unconscious. In M. Leuzinger-Bohleber, S. Arnold, and M. Solms (eds.), *The Unconscious: A Bridge Between Psychoanalysis and Cognitive Neuroscience*. London: Routledge, pp. 16–35.

Solms, M. (2020). Entrevista en "Recomendaciones neurocientíficas para los profesionales que practican el psicoanálisis". Seminario virtual de la IPA. Londres.

Solms, M. and Gamwell, L. (2006). *From Neurology to Psychoanalysis. Sigmund Freud's Neurological Drawings and Diagrams of the Mind.* New York: Binghamton University.

Solms, M. and Turnbull, O. (2011). ¿Qué es neuropsicoanálisis? *Revista Neuropsicoanálisis,* 13 (2). Depto. De Psicología. Universidad Cape Town, Sudáfrica, 133–146.

Solms, M. and Turnbull, O. (2002). *The Brain and the Inner World: An Introduction to the Neuroscience of Subjective Experience.* New York: Other Press.

Stagnaro, J. (2009). "Psiquiatría y neurobiología: el arte de curar y la ciencia del cerebro en crisis paradigmática". In *Jacques Lacan y los matemáticos, los lógicos y los científicos.* Buenos Aires: Escuela Freudiana de Buenos Aires.

Sulloway, F. (1992). *Freud, Biologist of the Mind: Beyond the Psychoanalytic Legend.* Boston, MA: Harvard University Press.

Talvitie, T. (2009). *Freudian Unconscious and Cognitive Neuroscience. From Unconscious Fantasies to Neural Algorithms.* London: Karnac.

Vanheule, S. (2011). Lacan's construction and deconstruction of the double-mirror device. *Frontiers in Psychology,* 2, (209). doi:10.3389/fpsyg.2011.00209.

Voos, D. (2013). Búsqueda del trastorno en el inconsciente. In *Mente y cerebro. El legado de Freud.* La neurociencia demuestra la eficacia del psicoanálisis. *Investigación y ciencia,* (62), 22–25.

Wallerstein, R. (2004). Introducción a la mesa redonda sobre psicoanálisis y psicoterapia. La relación entre el psicoanálisis y la psicoterapia. Problemas actuales. In *Revista Lacaniana. Las prácticas de la escucha y sus argumentos* (2). Buenos Aires: EOL.

Yellati, N. (2018). *Lo que el psicoanálisis enseña a las neurociencias.* Buenos Aires: Grama.

Yellati, N. (2021). Lo que el psicoanálisis enseña a las neurociencias. Conferencia dictada por modalidad virtual a través de Yoica AC. Accessed 24 August 2021 from: https://youtu.be/O22TlWW9bLA

Yue, G. and Cole, K. (1992). Strength increases from the motor program: comparison of training with maximal voluntary and imagined muscle contractions. *J. Neurophysiologie,* 67, 114–123

Zack, O. (2008). Mesa de apertura. *Discurso de apertura de RedAcción* (27). EOL.

Zack, O. (2016). *Vigencia de las neurosis.* Olivos: Grama.

Chapter 2

The intention in Freud's saying

There is no relationship between the unconscious and the brain

Beginnings of the relationship with science[1]

Freud was born in 1856 in Freiberg, a small and quiet town in the southeast of the principality of Moravia. Moravia was a small town in the northwestern part of the Austro-Hungarian Empire (now Příbor, Czech Republic), near the borders of Silesia. This empire had Austria, Hungary, and Germany as allies, united in an anti-Semitic ideology. Jewish persecution reached the Freud family – made up of immigrants, merchants, and Jews – especially the paternal family that had been persecuted for decades (Freud, 2006a; Jones, 1981; Soria, 2020). The ways for a Jew to be accepted were fourfold: slave-military, national German, noble, or cultural hero.

Knowledge was the ideal of the age. So, the Freud family bet on Sigmund, the only male child, raised with great responsibility and the soul of a conqueror. His mother was 21 years old when she gave birth to Sigmund, her firstborn. She then had five daughters and two more sons. One was Julius, who died at 8 months old while Sigmund was 19 months old; the other was Alexander, ten years younger than Sigmund. The family "placed in him [in Sigmund] all the hopes, those great hopes that Jewish families are pleased to forge with respect to their sons; illusions that among the Jews of Vienna, with their newly recognized rights, were, perhaps, especially lofty" (Jones, 1981, p. 8).

Jakob, the father of Sigmund, was not the type of stern father so common at that time. He was a wool merchant, interested in the intelligentsia and freethinking philosophies. His work began to decline just before Sigmund's birth. The introduction of machinery in Central Europe meant a threat to manual labor, producing considerable unemployment. "The inflation that followed the Restoration of 1851 further accentuated poverty in the city, which in 1859, the year of the Italo-Austrian war, was financially ruined" (p. 37). In addition, one of the consequences of the 1848 revolution was to make Czech nationalism a powerful factor in Austrian politics, thus stimulating the hatred of the Czechs against the Austro-German population, specially the ruling class of Bohemia and Moravia. Very soon this hatred "turned against the Jews [...] and [...] against the Jewish textile manufacturers" (p. 37), among them Jakob, who began to be considered as the culprits of this economic situation.

DOI: 10.4324/9781003458470-3

Freiberg, already a remote and decaying city, became a hostile environment. Because of this, the Freud family, of eight people, moved to Leipzig in October 1859, where they stayed for a year, before passing through Breslau, and finally arriving in Vienna, to one of the poorest neighborhoods, called Leopoldstadt, where other Eastern Jews from Hungary, Moravia, Bohemia and Galicia had settled (Teboul, 2019). There the family begins to grow. Jakob receives financial support from his wife's family. They stay in a three-bedroom apartment where Sigmund lives from the age of 4 "until he became a hospital intern" (Jones, 1981, p. 43). He goes to school (private education) where he was at the top of his class for seven years. Cabinets full of books begin to appear in his bedroom, "he even used to have dinner in his room, so as not to take any time away from his studies" (p. 43).

Reaching the end of secondary education, Jakob supported Sigmund in choosing a profession he liked and left him free to resolve this issue (Jones, 1981; Teboul, 2019). At that time, university courses were divided into two paths. On the left side was merchant or artist; on the right side was lawyer or doctor. Sigmund had no particular predilection for the status or activity of the doctor, he did not feel an attraction to medicine. "He did not hide, years later, the fact that he did not feel comfortable in the medical profession" (Jones, 1981, p. 51). He was moved by a desire to know about human life, but not to natural objects (Teboul, 2019). He was attracted to Darwin's theories, but he was not interested in the observation of physical phenomena.

> Although we lived in far from comfortable circumstances, my father insisted that, in choosing my career, I follow only my own inclinations. Neither at that time, nor later, by the way, have I felt any special predilection for a career as a doctor. I felt moved rather by a kind of curiosity, which was directed, however, rather to human affairs, than to the objects of nature. Nor had he come to grasp observation as the surest means of satisfying that curiosity.
>
> (Freud, cit. Jones, 1981, p. 52)

It was very that Freud was not interested in the observation of physical phenomena. Then what made him decide to study medicine? The "fact of having heard Goethe's beautiful essay on Nature, read aloud at a popular lecture by Professor Caryl Brühl, just before leaving school" (p. 52). "Goethe's dithyrambic essay is a romantic picture of Nature as a generous mother who grants her favorite children the privilege of exploring her secrets" (p. 54). That hidden praise of a God in the lyrical composition of the essay aroused a desire to access the privilege of exploring secrets, the complete opposite of what is verified, seen, and proven. In addition, his father read the Bible in Hebrew. Freud was familiar with the biblical account which contributed to the orientation of his interest.

The city of Vienna was third in Europe in terms of the production of knowledge. Freud circulated, since childhood, in this "nerve center" of the hegemonic culture of the world. He had in his hands the possibility of saving his family from anti-Semitic persecution. For this reason, as Jones (1981) indicates, no option was better than taking the path of science.

> After forty-one years of medical activity, my self-knowledge tells me that I have never been a doctor in the true sense of the word. I have become a doctor by being forced to deviate from my original purpose.
>
> (Freud, cit. Jones, 1981, p. 52)

That original purpose was the desire to help suffering humanity, to deal with the enigmas of human existence. That is to say, Freud pointed to the enigma, to what resists scientific proof: "In my youth I had felt the irrepressible need to understand something about the enigmas of the world in which we live and to contribute something, perhaps, to their solution" (p. 53). Freud felt imprisoned for devoting himself unconditionally to a discipline that was going to clash with his true interest. However, "the need for subjection offered by a discipline of a scientific nature, had as its outcome the triumph of the latter" (p. 57).

So, in 1873, at the age of 17, he entered a medical career at the University of Vienna. His unorthodox style, his initial doubts, and his way of being difficult to govern, lead him to take three years longer than expected. In addition, at the university it was a path of disappointment because he was subjected to the idea of, as a Jew, feeling inferior and ashamed of his origin or, as was said, of his race, and not part of the community of the people (Jones, 1981; Teboul, 2019). In Austria, Jewish persecution was daily, but Freud asserts his identity as a Viennese Jew and bears that name, resisting its contemptuous meaning. "He felt Jewish to the depths of his being" (Jones, 1981, p. 48). "My language is German. My culture, my realization is German. I considered myself a German intellectual, until I noticed the growth of anti-Semitic prejudice in Germany and Austria. Since then I prefer to consider myself a Jew" (Freud, 2005, p. 6). This is a life lesson: orthodoxies are fought by not rejecting the difference that inhabits every human being.

Freud graduated as a doctor on March 30, 1881. During the course of his degree he had worked at the Institute of Physiology of Professor Von Brücke, a professor with whom he had an important emotional bond. This professor suggested that he leave the Institute because the task of a researcher would not improve his economic position and to obtain a position as a well-paid professor he would have to wait many years. With the prospect of starting a family, this situation made him uncomfortable. So, following his teacher's advice, Freud resigns from the Institute he loved so much to enter the medical profession. Parental help – which was little because his father was already 67 years old, with the burden of a family of seven children and a uncertain financial situation – some small fees for his publications, a loan from some friends, among them Breuer – which he always returned scrupulously – and in 1879 a scholarship from the university, were his only income.

Freud was 26 years old and found himself without a livelihood. He did not want to be a doctor, but the situation was difficult, until he decided to take the path of this profession. For that, he had to acquire some clinical experience in caring for the sick, which is why he began working at the General Hospital of Vienna in 1882. He lives there for 3 years. He gathers experience while working incessantly on publications surrounded by the teachers who were outlining his training as a researcher. In March 1885, he applies for a travel grant offered by the University Fund of the Faculty of Medicine in Vienna to continue his neuropathological studies. Said scholarship consisted of remuneration and accommodation for six months at the Salpêtrière sanatorium. Due to his work as a researcher and professor – at that time he was already associated with the neuropathology chair – he had a chance of winning it. Indeed, he was distinguished as the candidate with the best prospects and he traveled to Paris in October 1885. There he meets Jean-Martin Charcot, responsible for the sanatorium, who extended his hand to continue studying the enigmatic disease of the time: hysteria.

Soul surgery

At the end of this formative experience Freud writes a report on his studies carried out in the sanatorium where he gives an account of the turn in his scientific interest: he goes from neurology to psychology. When he arrived in Paris his scientific interest was the anatomy of the nervous system. When leaving that city his spirit was persuaded by the problems of hysteria. Very early Freud began to turn his back on neurology while Charcot, his teacher, devoted himself to the pathological foundations of nervous diseases.

Charcot was an eminent French neurologist who brought about important changes in the conception and in the treatment of hysteria. Before Charcot, hysteria was considered a matter of the imagination, not worth occupying the time of a respectable physician. It was associated with demon possession and witchcraft, for which hysterical subjects were burned at the stake or exorcised at the time of the inquisition. Was considered a deviation treated by orthodox electrotherapy or a disorder of the uterus affecting women treated by removal of the clitoris. Hysteria was considered "the black beast", something that had to be defended against (Freud, 2011a, 2011f; Jones, 1981; Zack, 2016; Balzarini, 2023b). From Charcot hysteria acquires scientific status. Charcot rescues this affection from that medieval reading and turns it "into a disease of the nervous system, entirely respectable" (Jones, 1981, p. 231). Charcot's thesis was that organic motor paralysis was caused by some alteration in the set of relationships between various components of brain anatomy. This breaks with the theological, magical, obscurantist explanation and places hysteria in the paradigm of science.

Let us remember that the art of healing was from the beginning in the hands of the priests. What Charcot does is decenter the hysteria of the miraculous cure by theistic beliefs and transfer religious faith to the personality of the doctor, who begins to surround himself with a halo of prestige coming from that transferred divine power. Charcot stays on a level with Pinel who in the previous century, and also at the Salpêtrière, had freed the insane from their chains.

We are at the end of the 19th century, the decline of the Victorian era, "signed by the prevalence of ideals that operated in a repressive manner" (Zack, 2016, p. 71), the moment where a fruitful encounter takes place between this rare clinical entity called hysteria and a young Viennese doctor named Freud. Freud allowed himself to be captured by the secret that the hysteric woman carried, turning her symptom into an enigma that contained a message. This led to the assumption that the symptom speaks, that is to say that the neurotic symptom is a discourse. "Making it bear a message is realizing that its materiality is made up of words" (p. 75) and not neurons. Advancing in this sense in 1888 Freud very clearly says: "Hysteria is a neurosis in the strictest sense of the term; that is, no perceptible [anatomical] alterations of the nervous system have been found for this disease" (2011a, p. 45). Due to this new idea, his relationship with Charcot begins to be forced, until "in a note to the Poliklinische Vorträge in 1892 (page 100), Freud sharply criticized Charcot's theory as entirely inadequate" (Jones, 1981, p. 235).

Freud's desire to know led him to abandon his status as a neurologist, to shed the arrogance that the established knowledge gave him to defend the dignity of the hysteric's speech as had never been done before. It is necessary to underline this ethical position whose principle was to have separated from neurology. This is how psychoanalysis began: "[...] in the humble Viennese office of a neurologist concerned about the therapeutic failures of his discipline [...]" (Arenas, 2018, p. 119).

Freud had teachers who were not doctors but hysterics. What the hysteric teaches him is that psychoanalysis originates from neurology's denial that disturbances in sexual life are the essential factor in the etiology of psychoneuroses; that they constitute the invariable causes of neurotic affections and this, or rather the far-reaching conclusions that Freud will derive from it, is something that brain science, which dominated the eminent physicians of the day, including Charcot and Breuer, could not digest (Jones, 1981).

Thus, very early Freud distanced himself from the cerebral cause of hysteria: "Hysterical paralysis does not take the anatomical building of the nervous system into consideration at all" (2011a, p. 50). That is, "[...] hysterical conditions in no way offer a reflection of the anatomical constellation of the nervous system" (p. 53). Indeed, "[...] the idea that there would be a possible organic disturbance at the base of hysteria must be rejected, and it is not lawful to invoke vasomotor influences (vascular spasms) as a cause of hysterical disturbances" (p. 54). Freud exposes from the outset his break with localizationism; he demonstrates, already in his first publications prior to the establishment of psychoanalysis, that his thesis did not consist in defending material causality.

These "Freudian ideas rejected the scientific assumptions from which German medicine had made its great advances" (Jones, 1981, p. 10). For the disciples of the Helmoltz School, "the idea that the brain or nervous system was not the cause of a malfunction of the body was worse than professional heresy" (p. 10). Freud had been educated in the tradition of those disciples, and he was expected to continue and honor it, but this humble Jew has the courage to go against the thinking of the time, break the establishment and pierce science.

> [...] in the past, the other side of this relationship, the action of the soul on the body, found little favor in the eyes of doctors. They seemed to fear that if they gave some autonomy to mental life, they would lose their footing on the safe ground of science [...] the study of the brain and nerves of patients of this class has not allowed us to discover any visible alteration up to now, and even many features of their pathological picture dissuade us from expecting that one day, with finer means of examination, we would be able to verify alterations capable of causing the disease.
>
> (Freud, 2011i, pp. 116–117)

Even with more refined means of examination, such as modern neuroimaging techniques, psychic causality would not be clarified. This implies that psychoanalysis, from its beginnings, is a practice of words, not of images. While neuropsychoanalysis is based on a rigorous statement that postulates that what is visible is scientific and that what is not sustained in some physical reality is not true.

Let's put the question here: can the existence of intangible realities be proven? In the year 2000, the movie *The Matrix* points to this and proposes two scenarios. On the one hand, the physical reality, located in the cabin of the ship where we see the body of Morpheus as pure biological support, tied up and alienated from his own domain. On the other, virtual reality, a space without physical supports where the fate of every subject is played, from which the subjected body knows nothing, except for the effects it feels, just as it happens to Morpheus and Neo. Virtual reality has such a consistency that Morpheus makes blood come out of his nose (Pulice, Zelis, and Manson, 2019). This is a way of representing the proof of intangible realities.

From a conception that starts from the brain to another that excludes it

In 1888 Freud published the first two parts of the writing "Some considerations with a view to a comparative study of organic and hysterical motor paralysis". Both parts develop the neurological bases of organic paralysis. This was followed by five years of complete silence until the last decidedly psychoanalytic part was published in 1893. The causes of these five years of delay are, says Freud (2011j), accidental and personal. We don't know more than that. But we can venture a possible explanation. It should be known that this work occupies a position in the "watershed" of Freud's neurological and psychological writings; it represents the criticism of the school of his former teacher Meynert who maintained that the cerebral cortex is a mirror of the parts of our body (Jones, 1981); it represents the separation with Charcot and the abandonment of hypnosis. These landslides were not easy.

From this study it turns out that Freud cannot hide his position: "it is evidently impossible that this anatomy can explain the distinctive features of hysterical paralysis. For this reason, it is not lawful to draw conclusions based on the symptoms of these paralyzes regarding brain anatomy" (2011j, p. 205). Freud drops the cerebral cause of hysterical symptoms for which this article on motor paralysis is crucial because it goes from a conception of paralysis that starts from the brain in its materiality to a conception that excludes it (Yellati, 2021). It is a rupture in its beginnings, a very early proof of Freud's intention.

Difficulties in the project

In 1895 Freud was devoted to the theoretical problem of the relationship between neurology and psychology. His reflections lead to the unfinished manuscript "Project for a psychology for neurologists". Parts I and III are essentially theoretical, while Part II presents clinical material linked to psychopathology, in that material sexuality occupies a prominent place, but in the theoretical parts it has little role. This awkward divorce between the clinical and theoretical significance of sexuality is further proof of the non-relationship between neurology and sexuality.

Indeed, there is a theoretical-clinical difficulty that exists in Freud's intention to say: not all its quantity can be discharged. The exogenist Freud, who describes a receptive apparatus insofar as he places all the emphasis on the effect of the environment of the organism and on its reaction to it, recognizes the existence of endogenous excitations and, when he barely examines their nature, he finds that endogenous stimuli place the individual under the pressure of life. The individual cannot escape from these internal stimuli and, therefore, they produce constant stimulation: "The organism cannot withdraw from these stimuli as from external stimuli, it cannot apply its Q to flee from the stimulus" (Freud, 2011k, p. 341). If there is no cancellation for that stimulation, which comes from the body, then the flight reaction, the reflex act, is not enough as a defense to free oneself from the displeasure produced by the tension charge. Another proof of the lack of relationship between the nervous system and mental life.

Up to this point, Freud still does not have the concept of the drive, he does not have a way to name this difficulty that he encounters in this localization project. He has to abandon the principle of inertia and present the principle of constancy. It will say that the psychic apparatus cannot discharge the voltage to zero, but it will try to keep it as low as possible. There is then something inevacuable in human sexuality that poses difficulties in the effort to pass psychic causality through the paradigm of quantity.

However, Freud tries not to abandon the quantitative paradigm and makes notable theoretical efforts. He begins by proposing two classes of neurons. In the first place, those that allow the amount of excitation produced by a stimulus to pass through, as if they had no barrier-contact and therefore, after each excitatory course, remain in the same state as before. Second, those whose contact barriers are reinforced in such a way that the excitatory quantity can only pass through with difficulty or only partially. Freud calls the first passenger neurons, those that do not oppose resistance and do not retain, they only serve perception. He calls the non-transient second neurons, afflicted by resistance and retainers of quantity, carriers of memory. He designates passing neurons with the Greek letter phi (Φ) and non-passing neurons with the Greek letter psi (Ψ).

The psi neurons are those dedicated to inscribing reality and constituting memory. What produces a lasting alteration given by the excitatory course is inscribed. We could say that the contact barriers, which will later be known under the concept of synapses introduced by Foster and Sherrington in 1897, two years after Freud wrote this, are altered and susceptible to higher levels of conduction and less impasses, and therefore more similar to the fi system, which produces higher levels of facilitation of impulse conduction.

For this Freud, memory is constituted by facilitations. When there are facilitations, there is an experience that alters the normal state of retention of these neurons and also the state of the contact barriers. That experience, which produces alteration, passes into memory. Not anything can alter the facilitations between psi neurons. What has that value then remains constituting memory. In this sense, memory is the power of an experience to continue producing effects over time and quantity is replaced by facilitations. The facilitations come to replace the excitatory course produced by the magnitude of the impression. So the facilitations serve the primary function, which is the discharge. In its essence, neurons maintain an eagerness to discharge quantity. This is the natural point of view, the primary function.

But there is a type of neuron, says Freud, that does not have these characteristics, the psi. They have them when an experience exceeds their thresholds. There it returns to its natural state, giving free rein to the excitatory process, preventing the quantity from being reduced. So what caused a group of neurons to break out of their natural state and start to hold the course of excitation? The rush of life. The urgency of life, Freud says, is the cause of some neurons passing from their natural state in which they seek the discharge of excitation to a retentive state that collects a part of the excitation.

In short, the biological point of view is the receptive and reactive nervous apparatus, but Freud, with the concepts that he builds, indicates that it is necessary to get out of that point of view. The apparatus receives stimuli from the outside world and reacts sending the discharge toward motility, toward action, toward a state of peace for the individual. The point, says Freud, is that in this tendency toward motility, all possibility of quality is lost, and all possibility of giving particular value to that quantity is lost. "The discharge, like all, goes to the side of motility, as a result of which it should be pointed out that in the motor circulation all quality character is lost, all particularity of the period" (p. 356).

The organism seeks to balance itself, as we have read from these neurosciences when they speak of the unconscious from the automated or dopaminergic point of view. It was necessary for Freud to organize a theoretical approach divided into two systems of neurons to understand these laws of reception–reaction. However, he says that this separation into two classes of neurons has no histological or morphological foundations.

> It will be objected to our hypothesis of contact-barriers that we assume two classes of neurons with a fundamental diversity in their conditions of function, for whose separation, at first sight, all other grounds are lacking. At least morphologically (i.e., histologically), nothing is known to support this separation.
>
> (pp. 346–347)

If there are no biological foundations that support the hypothesis of the separation into two classes of neuronal functions, a hypothesis that gave rise to this initial effort by Freud, there is no relationship between nervousness and meaning.

Freud then has a problem. He wonders where the quality of the things that arise in our consciousness comes from. He says that it cannot be from the perceptive pole because the quality obeys a higher process than perception; nor can it come from the external world because there are only physical objects in motion that are not valued therefore the quality does not come from the system fi (Φ); it cannot come from a process of remembering because in itself the memory does not contain quality, therefore the quality does not come from the psi (Ψ) system either. So he must add a third system of neurons, which is designated with the Greek letter omega (ω).

The omega system is excited immediately after perception, but not upon memory, as the psi system would be. Here we have a leap from quantity to quality. With the addition of this third system of neurons, he adds the paradigm of quality. It is a combination that tries to explain the way in which the psychic apparatus changes the external quantity into quality and in this way that the apparatus that deals with separating the quantity can triumph. The omega system gives quality to send it to consciousness. It works when the quantities, smaller than those occupied by the psi system, are disconnected. In other words, the ω neurons, upon receiving a smaller excitatory quantity, become more impassable, more impenetrable than the psi, for this reason they can achieve the goal of qualifying, because they can retain more than the psi. The psi neuron system has the impasse condition, but it becomes passable if the amount of excitation is sufficient to produce facilitations: "We saw that the pass condition depends on the interference of Qn, the Ψ neurons are already impasse" (p. 354). Now, the more impassable, the more impenetrable are the neurons of the ω system that receive a smaller amount. But also the ω neurons are invested with quantity, therefore there is not a total cancellation of the quantities.

These entanglements of the groups of neurons, the different quantities (Qn), show a Freud forced into a commitment to science. The quantity problem is verified at each theoretical step. Already in its title, which was not in the original manuscript, there is something that comes from no one knows where, without a subject, because it does not begin by saying "My project", or "The project of", it starts by saying "Project", just plain. The subject is removed, it is not known who it comes from, it is not known who it belongs to, the subject is not there. The title is indicated as addressed to others, "for neurologists", an obstinacy to impress them or to please the master? And it is that in order for it to be scientific and obtain recognition in the field of the ideal of the time, the project has to "deal with something material" (Miller, 2015a, p. 178). Even today, this correlation between science and matter is being discussed.

Now, if Freud had a scientific project on his hands, why was it published 55 years after it was written? Strachey tells us that he wrote it at a hurried pace, but, while writing it, he was the subject of the severest criticism. "In later periods of his life he seems to have forgotten about it, or at least he never mentioned it. And when in his old age they put it back in his hands, he did everything possible to destroy it" (cit. Freud, 2011k, p. 333). What happened between Freud and Project? Why was it the object of such severe criticism? What value did it contain, if not the lack of relationship between two substances whose relationship it had to necessarily demonstrate? Trapped, consumed, he couldn't solve it, he couldn't close it. This is what he tells Fliess: "Scientifically I am doing badly, so stubborn in 'psychology for neurologists' that it regularly devours me whole until I have to interrupt really tired. I have never gone through such extreme concern" (1986a, pp. 129–130). Years later this exhaustion again:

> As for the mental apparatus, it will soon become clear what it is. And please don't ask me about the material it's made of. Psychology is not interested, it may be as indifferent as for optics to know whether the walls of the telescope are made of metal or cardboard.
>
> (2006b, p. 181)

Strachey reminds us that "[…] Freud himself ultimately discarded the entire neurological frame of reference" (cit. Freud, 2011k, p. 336) for which reason this manuscript should continue to be considered as "an unfinished draft, unauthorized by its creator" (p. 336).

The last man

As we have been reading, Freud was trying to build a scientific explanation for the psychic, but at odds with the intention, with the repressed nature of his saying. In much of the correspondence with Fliess, Freud expresses doubt about the biological causation of the neuroses. Fliess was the confidant that Freud could not lose. There the important phenomena of transference were masked, that is, the place of Other that Fliess occupied for Freud at a time when his discoveries began to cause revolutions in the field of science, he was increasingly subjected to a boycott that provoked in him challenging responses, feelings of hatred that needed a place to be housed, understood, and worked on. However, the differences were noticeable. Fliess was on firm ground in his knowledge of the nervous system while Freud doubted that neurotic manifestations were strictly determined by biological laws. The tension grew and the only thing they asked for was recognition of each other's advances and theories. "Neither of them could have really understood much of the other's work. The only thing they demanded of each other was each other's admiration for what each did" (Jones, 1981, p. 309). They tried to impress each other, to elevate themselves, exposing their efforts to arrive at good scientific form. Finally, the meetings and the letters ended in 1902, although the relationship had fallen apart some time before (Teboul, 2019; Jones, 1981). In 1897 Freud had already stated that theorizing about the inscription of subjective representations in neurons is a delirium:

> For science I am a lost man [...] I no longer understand the state of mind in which I incubated psychology; I don't conceive that he could have locked you in on you. I think you are too polite anyway, it seems to me like some kind of ingenious delirium.
>
> (Freud, 1986a, p. 159).

"And it is true, the project of a psychology for neurologists that wanted the representations of language, the subjective representations, to be inscribed in the neurons, that was a delusion" (Bassols, 2011a, p. 86). Although the ideas of this Project contain the germ of what will later be the basis of fundamental concepts in psychoanalysis, it must be assumed that such concepts came to light on the condition that Freud managed to separate himself from the neuronal paradigm. Although some scientists continue to be very kind to that dream of translation that goes hand in hand with the nightmare of quantitative ideology (Bassols, cit. Laurent, 2005).

Thus, Freud was caught in an imaginary relationship with Fliess, the Project, and neurology. He made himself come from these others a unifying image, a kind of romantic love that seeks to find in the other the reflection of oneself. If divergences came to light, unexpected tensions ensued. A relationship of pure resemblance, but not pierced. A love where the other is at the service of fulfilling all that I lack, but does not make room for what does not fill me, what decompletes me.

As Lacan (1988a) indicates, from an imaginary relationship it is possible to expect an aggressive outcome. The neighbor is whom the human being can satisfy his need to attack. Hegel (1966) spoke about the fight for recognition to the death that the subject must fight with the other for survival when he enters into rivalry. In such a dual relationship, a third party is excluded. A third element cannot enter between those two elements that are completely and perfectly identified. This third element is that would come to introduce something of the law, as Lacan (2009f) locates with respect to thirdness, from where the symbolic is inscribed, from where comes what it regulates. If that is excluded the relationship between the subject and the mirror is one of pure pleasure.

The mirror falls apart when the subject has to enter the culture. As Bassols (2011a) says, "the mirage in which cognitivism today is lost, as an ideology of neurosciences, is to assume, 'by elimination', that the seat of language [...] is the brain as its organ" (p. 93). Who can insure his? The "say it in your own words" invitation, who can say that the words are theirs? The story is full of allusions, blank spaces, puns, and misunderstandings, which reveal that there is another discourse: that of the unconscious (Castanet, 2023). However, neuroscientists strive to define it without falsehoods.

If the brain absorbs the unconscious, the world will shrink and the neuroscientist will be so big that he will be left alone. Nietzsche (cit. Han, 2022; cit. Simonet, 2019) said: "The earth, then, became small and on it wanders the last man who shrinks everything". The Nietzschean idea of the last man is embodied quite well, says Miller (2015a), by the neurocognitivist. It is the man who jumps alone, the man who knows everything, everything happens in this form that today is neuronal (Miller, 1978). The neural man remains alive after the elimination of everything, enjoying a majestic image until he dies. For this reason, says Miller (2016c), the advance of these neurosciences goes in the direction of the death drive.

Drive fixation does not take place in the brain

During the year 1896 Freud (2011g, 2011n, 2011p) was going to postulate the independent source of detachment of displeasure, independent of what would later be ordered in the phallic field. These are fixations that are prior to the word, prior to language, prior to the Other, which do not form a chain. As Delgado (2021) indicates, this is something that was not symbolized and that will give the symptomatic remains that remain after an analysis. Something that causes neuroses and that is not tied to the meaning that, in its permanent return, evidences its isolation from the chain of signification.

In December 1896 Freud sent Fliess the famous "Letter 52". From this we know that the detachment of displeasure is characteristic of the translation from one phase to the other while something is left out, something is not translated. Pedagogical endeavors, consciousness-raising, and cognitive training work, present difficulties because the memory recovery process is never complete. Memory and consciousness are excluded (Freud, 2011g).

As Freud explains, the symptomatic remains that are impossible to cure are associated with a mark that pre-exists the memory traces, an inaccessible mark, a material that Freud designates as perceptions: "[...] in themselves they do not preserve any trace of what happened" (2011g, p. 275). Said mark is fixed with the first Sign of Perception, which is equivalent to the first transcription, even "completely insusceptible of conscience" (p. 275); it is not what is perceived as such, but rather what is transcribed from what is perceived. If it is not what is perceived, its originality cannot be recovered. Then, Freud says, there is a second transcription that is equivalent to the unconscious, where the fictions of what has been lived will be, but not yet accessible to consciousness. Only in a third transcription will it be associated with a word-representation.

This indicates that not everything that happened is transcribed, not everything that has been experienced is added to the chain of representation. As Delgado (2018) says, that unmarked mark, prior to the first transcription, has no relationship with the signifier, it does not respond to the unconscious, which is the second order. On this mark will mount the whole series of signifiers that structure the unconscious, but unattainable in its real state through language. There remains then an element that is displaced from the associative commerce, due to the effect of the primary repression that left that drive fixation without articulation. So fixation does not have its place in the brain.

Lacan comments on "Letter 52". He says that the phrase "layers of the organ" is analogous to this triple Freudian stratification. In the phenomenon of vision, light reaches various layers, producing refractions from which changes occur in vision, but not in the organ. That is, refraction is associated with vision, but not with anatomy. Thus, says Lacan (2013), it happens in the unconscious:

> I will mention again, for those who have already heard my lectures on the subject, letter fifty-two to Fliess, which comments on the scheme, later called, in the Traumdeutung, optical. This model represents a number of layers, permeable to something analogous to light, and whose refraction is supposed to change from layer to layer. That is the place where the issue of the subject of the unconscious comes into play. And it is not, says Freud, a spatial, anatomical place, for how else can we conceive it as it is presented to us?
>
> (p. 53)

When we deal with humans there is no memory photograph. The psychic apparatus is not a photographic camera that records moments and then reproduces them exactly as they happened. The sign of reality is lost from the referent episode that preserves its origin. However, neuroscience knows how this origin can be modified.

Of a non-biological body

The text on hysterical paralysis from the year 1893, which we have already presented, is the explicit demonstration of Freud's break with the biological causality of symptoms. There Freud says that hysterics do not verify the localizationist intention, they simulate other paralyzes that cannot be explained by anatomy. With this Freud discovers another body that is not that of anatomy, a body made by words. According to Yellati (2018), this early observation of Freud is a prelude to an obvious theoretical consequence: the organism is not the body.

This advances in 1905 when he affirms that sexuality is born propped up in certain bodily areas, that the goal of sexuality is satisfaction in those so-called erogenous areas, so there is no hierarchical area, and there is no governing area for satisfaction. Sexuality is opposed to development ordered by the end of reproduction. Thus, Freud demolishes the scientific myth that only those who reach puberty would be sexualized. If sexuality is separated from reproduction, we have another proof that separates the unconscious and biology.

In those years, Freud defines the child as polymorphously perverse, that is, the child presents the entire catalog of what could be considered perverse, amoral acts. Perverse, not as it is understood at this time in the sense of aberrant, monstrous. Perverse for Freud means outside the normal sexual pathway, satisfying oneself through partial areas of the body, for example, oral or anal. This discovery of infantile sexuality was experienced as an offense in Freud's time, an ancient time, of righteous and controlling passion. Until today, sexual practice in children (erections, masturbation, and actions similar to coitus) is taken, in the medical literature, as an exception, degeneration, or a horrific example of early corruption, just look at the lack of scientific consensus about the diagnostic criteria of paraphilic disorder. Human sexuality does not fit the criteria of normality.

In 1910 Freud discovers that an organ can stop being at the service of preserving life to be taken over by the drive. He takes the example of neurotic blindness and affirms that the blind person may not organically have vision, but he has the gaze available in his own phantom.

> Suitable experiments have shown that the hysterically blind see, however, in a certain sense, though not in the full sense [...] the hysterical blind are blind only for consciousness; in the unconscious they are seers. Experiences of this nature, precisely, were the ones that forced us to separate between conscious and unconscious mental processes.
>
> (2012a, p. 210)

What neurotic blindness demonstrates is that the vital functions so precious in the service of self-preservation, such as vision, can be disturbed as long as the function of the organ is erogenized (Yellati, 2017). If the organism is no longer solely at the service of preserving life, the basic assumption of biology falls and, once again, there is no parity between the organic and the unconscious.

Hippocrates had inaugurated for medical practice a sacred commitment to life in which the doctor took an oath that his action will always be aimed at maintaining life. That Hippocratic idea of perpetuating the species, of avoiding harm, is absolutely questioned. Indeed, "what Freud shows about the functioning of the unconscious has nothing biological" (Lacan, 2009b, p. 30).

To make a mistake is to say it right

> Misunderstanding has hitherto been the greatest force on earth
>
> Edith Södergran (1919)

Before Freud, failed operations had been explained in cognitive theories and had always been traced back to attention deficit, fatigue, or semantic emptiness. The word deficit refers to what is essentially physical. The deficit can be located in some areas. Instead, the equivocation is significant. The equivocations evidence what is not exhausted in the identification, what the identification does not fill. In such a way, Freud (2012d) relocated the misunderstandings to another field than the neural. He rescued failed operations from being understood as distraction, inattention, derived from fatigue, or "side effect of certain mild pathological states" (2011e, p. 170). Misunderstandings, starting with Freud, ask to speak, not to be corrected. "When things exist, even if they are a little crooked, it is not mandatory to straighten them" (Miller, 2004b, p. 14). It is not a sin to fail. On the contrary, to fail is to say good. Without fail, the truth could not be told because it is told halfway (Lacan, 2008e).

> Such phenomena are usually referred to pathology – as long as they are not completely ignored, as is the case with failed operations –, and every effort is made to give them physiological explanations, which have never been satisfactory.
>
> (Freud, 2011e, p. 170)

Errors are relevant to the psychoanalyst. After what a subject said wrong is the trace of the position that decides in whose substance the indeterminacy of the suffering is concealed. Indeterminacy because the subject does not know himself, the patient who errs does not know his subjective determination. When the neurotic speaks he errs the truth because he is divided. There are several anecdotes to illustrate this in a fun way. Let's see some.

> It was the moment when he gave Perón no chance to come back, Lanusse said on television. Lanusse is going to pay tribute to Sarmiento in Mendoza, which is broadcast on the national network. And Lanusse says on television for the entire country [Argentina]: "I come to honor Juan Domingo Sarmiento" [he meant Domingo Faustino Sarmiento, an Argentine hero who collaborated in public education, but he made a mistake and said "Juan Domingo Sarmiento", being Juan Domingo Perón, the name of Lanusse's political enemy], this of course goes through what is called a furcio on television. That is, a man who was in a violent dialectic with Perón, says "I have come to pay homage to Juan Domingo Perón". If someone asks him, did you want to honor Perón? "Nothing to do with it, I hate him, I want to kill him, and if I can, I'll seriously kill him," but he pays tribute to him. The clear division is this.
>
> (Chamorro, 2011, p. 108)

Another anecdote, from the year 2022, when the war between Russia and Ukraine broke out, former US President George W. Bush gave a conference in Dallas where he said: "The decision of one man to launch a totally unjustified and brutal invasion in Iraq [...] I mean, in Ukraine" (Bush, 2022, p. 1). In other words, he condemned Vladimir Putin's invasion of Iraq, when in fact he meant the invasion of Ukraine.

We have another accidental event in the speech that Alberto Fernández gave in Plaza de Mayo after learning of his victory in the presidential elections at the end of 2019. "Four years we heard that we would not return, but tonight we returned and we are going to be women" (Fernández, 2019, p. 1), harangued the president, although he quickly corrected his last word with the term "better".[2]

Do Lanusse, Bush, and Fernández have any neural or brain deficits? Well, there are those who would be tempted to say yes, but the point is that if psychoanalysis is driven by these accidents, neuroscience claims, success can never be assured.

That is why Freud says that failed operations "begin the destiny of psychoanalysis to assert itself in opposition to official science" (2011e, p. 173).

> If cognitive science investigates the functions that allow humans to know the world in which they live and the organic damage that prevents it, psychoanalysis starts from the failure of these functions to demonstrate that said human is a divided subject.
>
> (Yellati, 2018, p. 13)

Freud was in charge of demonstrating that the errors have another causality than the one that indicates the deficiencies to which attention or memory can succumb. The meaning is always equivocal. Psychoanalysis must make the neurosciences understand that they work by annulling the subject of the unconscious (Lacan, 2012f).

Scientific infatuation

In 1911 Freud was concerned about the dominance of science and wondered how psychoanalysis would survive in the face of the fury that there is to cure.

> The idea of healing in the psychic field is based on the notion that the psyche would be like an organ of the body and would be confused with the functioning of the brain. Now, psychoanalysis does not deal with the psyche, but deals with the unconscious, which is very different. The unconscious is not an organ. It does not perform any function of knowledge of the world and, in the field of the unconscious, healing does not make sense.
>
> (Miller, 1994b, p. 9)

No one is cured of their way of enjoying. Arrangements can be made to live a little better, but psychoanalysis does not pursue a cure. Freud (2012f) starts from this ethical position and asks: how does a psychoanalyst become? "I can start by saying that psychoanalysis is not the result of speculation, but the result of experience" (p. 211). That is, whoever intends to work as a psychoanalyst must be psychoanalyzed,

> [...] otherwise, would introduce into the analysis a new type of selection and distortion much more damaging than those caused by a strain of your conscious attention. For this [...] it is lawful to demand, rather, that he has undergone a psychoanalytic purification, and taken note of his own complexes that could disturb him to apprehend what the analyzed offers him.
>
> (2012e, p. 115)

It is convenient to review Freud's expression "analyzed". If the patient is analyzed then there is someone who analyzes another. Someone who analyzes would be, as Chamorro (2011) says, in a position of master, that is, in a position of subject because he begins to exert a meaning, he begins to say "what happens to you is this and this", "you say this because what you want is this and this", "what you must do is this, because it is good". And if things don't work out, he's going to accuse the patient of not being a good patient.

> When the patient wants to leave the analysis, because things don't work out and because the analyst is blaming him for not working, because he doesn't work well, because he was absent once twenty years ago – there is always a reason to blame – then the patient finally says "I'm leaving here", and, when he leaves, the master says "See, this proves what I was saying, and besides, life is going to go very badly, suicide is likely", I heard it.
>
> (p. 174)

The analyst's position is not that of master, but, on the contrary, that of object cause. The analyst causes the analysand to speak, in the direction of the object that implicates him, which is embodied by the analyst. If the analyst is not in this position what happens is a cure directed by the illusion of healing, which is not an analysis, but rather a practice of suggestion. For this reason, for Lacan, a psychoanalysand is not a psychoanalyzed, he is not in a passive position being analyzed by someone, but the analytic act also concerns him; he analyzes himself, with someone he chooses and who he actively works with. And who works pays, pays with his knowledge, gives his consent to knowledge, yields the involvement of his unconscious. It is the inverse logic of the laws of the market (Sinatra, 2017).

Today the market offers treatments that, far from opening any questions, give slogans that teach us to reject the impossible to bear. What is real is what is opaque to meaning, so it cannot be treated with tips to manage emotions, as is now heard in various practices such as biodecoding, family constellations, coaching, or mindfulness. The rules that are given to be happy fail. The analyst allows with his desire a place to house this unbearable real every time the subject encounters it again.

We know that science is divided into specialties. Each specialty has its own concepts that allow this special group to differentiate itself. The specialist, therefore, knows *a priori* what special material must be attended to. As Chamorro (2011) says, the specialist makes sense of things with the structure of the diagnosis where he has all the symptoms. "Tell me what is happening to you", the subject begins to tell some elements and they are immediately located in the classification structure, "this happens to you too", he begins to orient the symptoms and investigates them anticipating that he has them. Knowledge is not on the side of the analysand, but on the one who is supposed to know. For this reason, various specialists must be visited. But if knowledge is on the side of the analysand, there is only one analysis.

The specialties are microworlds. As Miller (2023) points out, a family is a microworld and has its own vocabulary, signifiers with peculiar meanings only valid within said community. The same phenomenon occurs within a group of friends, sometimes with the voluntary fabrication of their own signs, of ritualized practices to distinguish themselves from other gangs. Each microworld has its jargon. Likewise, every profession has a specialized language, closed to others, and the more scientific, the more stereotyped its language. "Every specific field of knowledge has its own nomenclature and, for any layman, this nomenclature is usually enigmatic" (Goya, 2017, p. 45). The point is that we don't perceive what someone is trying to say when they use the same words as ourselves. So, understanding yourself too much is not understanding yourself. That is the disease of the microworld, that the repetition of the statements erases the enunciation of another (Miller, 2023).

For example, in a case presented by Sanchez and Sanchez (2004), phobia is associated with a predisposition in the processing of information referring to the brain nucleus that is responsible for emitting threat signals. The condition, of course, is evaluated by means of questionnaires with potentially phobic situations, that is, questionnaires aimed at the patient confirming the phobia. With questions like "Does it make you anxious to see a bird in a cage?" and boxes "A little, never, sometimes, frequently" we proceed to establish a phobic scale that goes from 0 to 100 that does without the testimony of the subject. The subject is not heard in what is unique about it. It is true that he is allowed to speak, but "whatever his answer may be, it will necessarily be comparable to that of another […] It will be in the average or below" (Miller, 2004c, p. 158). You can do something unique, like tear the sheet or not respond, but at that moment you will be in the percentage of refractory subjects, that is to say, you can manifest yourself, but you'll be considered a misfit.

In turn, the phobia is treated with special techniques: distraction, cards with a list of positive memories, change of catastrophic thoughts, and some advance preparation where the sufferer is exposed to the phobic situation that inoculates him to stress and adapts him to the management of a simple phobia. It is not a matter of knowing what the symptom means, nor of knowing its cause, but rather that knowledge dominates jouissance while the emotions remain contained in that ultra-reduced artifact of knowledge by the scientist.

If we follow Freud, phobia belongs to the register that we call significant, while fear is a pathetic affect that is experienced, and felt. The phobia is a formation of the unconscious, but it is not a fear, it is a lucubration of knowing about fear, it is its signifying armor (Miller, 2017a). Psychoanalysis does not teach the subject to launch without fear. We are not there to enable you to overcome your fears. On the contrary, we ask him to expel knowledge, to empty himself inside, to purify himself of the waste that it contains, which places knowledge on the side of the subject.

So there are at least two therapies: one that treats the real and another that corrects it; one that is oriented by the real and another that denies the real or seeks to adapt it. On the side of those who seek to adapt it, Wallerstein (2004) thinks about whether "the patient should adapt to the treatment or the treatment to the patient" (p. 174). Adaptation or adequacy is on the path of correction. The psychoanalysis is contrary to any application of correction procedures, it does not try to adapt the subject, but rather to produce the subject's agreement with himself (Laurent, 2006).

The progress of psychoanalysis is further delayed by the terror felt by the ordinary observer to see himself reflected in his own mirror. Scientists usually deal with emotional resistance with arguments, and are thus fully satisfied! Whoever wishes not to overlook a truth will do well to distrust its antipathies, and if he intends to critically examine the theory of psychoanalysis, he must first analyze himself before doing so.

(Freud, 2012f, p. 215)

It is a crucial indication. Psychoanalysis will not progress, but not by not allying with biology, it will not progress if the person who practices it does not want to know or does not want to be warned about the relationship with his own unconscious. This is what Freud tries to say in the preceding quote when he is addressing the scientists who tried to start from the psychoanalysis concepts to obtain individual recognition. He warns them that they will not get far if they try to satisfy themselves with scientific explanations that do nothing more than feed their own image by reciting a purely university discourse while not using personal psychoanalysis as the foundation of their training and practice. They will be able to contribute to the cause of psychoanalysis as long as they are aware of the relationship with their own reality. Otherwise, Freud indicates in 1932, it will be the scientific explanation that governs his practice:

[...] the scientific worldview already distances itself remarkably from our definition [...] He asserts that there is no other source to know the universe than the intellectual elaboration of carefully verified observations [...] It was reserved for our century to discover the presumptuous argument that such a worldview is as poor as it is heartbreaking, that it neglects the demands of the spirit and the needs of the human soul.

(2006f, p. 147)

Psychoanalysis may question social phenomena, but it is not a worldview. It is an absolutely original pragmatic, drawing on various readings, but without being confused with them. Freud knew that psychoanalysis was not accepted by scientists who claimed to defend it. But that didn't bother him as much as that some scientists ripped concepts from psychoanalysis to confuse them for the whim of making contributions to their discipline. He asks them to stop:

It is new, yes, that in scientific society a sort of bumper has been formed between analysis and its opponents, people who accept some of the analysis and even declare themselves to be its supporters under hilarious restrictive clauses, but instead disavow another part, something they never consider to have proclaimed loud enough [...] These eclectics do not seem to care that the edifice of psychoanalysis, although unfinished, still constitutes a unit from which anyone cannot tear elements at whim [...] He excuses them, it is true, that their time and their interest are claimed by other things, namely, those for whose domain they have made such valuable contributions. But shouldn't they suspend their judgment, instead of so decisively taking sides?

(2006e, p. 128)

When one takes sides so decisively one becomes easily a prophet of an ideological party from which the results are pushed toward a universal metaphor that, grandly, can explain a whole number of phenomena in the world. It is there, says Laurent (2020b), that a metaphor begins to impose itself with its certainty. And an ideologue is not a scientist.

> What neurosciences do is very good. Neurosciences, when they are science, are excellent. In short, they teach us things, although they take a while to start making them, but when they go to make prostheses, which allow paralyzed people to read the orders of the brain, which allows them to articulate an exoskeleton, excellent. We all have an acquaintance in our ambit who could well take advantage of this. So, as a science, there is no problem. It is when a scientific ideology leads that it is a problem. Because science is always accompanied by a scientific ideology. When physics emerged in the world in the 17th century, an idea of the universe as a great clock, as a great physical machine, was immediately established. But, that was an illusion […] When biology arose, in its first results the world was seen as a living animal. Well, bad metaphor, the world does not obey the imperatives of the living being […] Those moments in which a metaphor prevails with its certainties.
>
> (p. 74)

Those moments are when all knowledge is rejected, when the primacy of the pleasure principle is upheld and "the dimension of the real as what truly governs the dialectic of discourse is rejected, it is dangerously close to psychosis" (Barros, 2004, p. 27). Lacan in the 1960s highlighted the relationship between psychosis and science, since they have something in common: the rejection of all knowledge (Laurent, 1991). It is about the "[…] virus that today motivates the absolutely delirious epidemic of wanting to represent and map all the subjective functions in the brain" (Bassols, 2011a, p. 78).

> Finally, we will develop a new perspective on the relationship between the conscious and the unconscious through an examination of the transactions between the core and the multitude of functionally isolated routines that are related to automatic and unconscious processes. This test clarifies the role of consciousness in learning and memory. Taken together, all of this effort suggests that it may be possible to untie the knot of the world after all.
>
> (Edelman and Tononi, 2002, p. 96)

"It is possible to untie the knot of the world" is clearer when it is translated: "We have invented happiness". It is shown that this machine for evaluating the effectiveness of psychoanalysis leads to madness. Freud teaches us that with delusion a subject can be stabilized. If it decompensates, they will produce the double, the triple. There will be no way to dismantle its axioms, this is staying away from the experiences of psychoanalysis, that is to say the "abandonment of the unconscious" (Peteiro, cit. Bassols, 2011a, p. 213).

> Excuse me, then, if I don't continue down that path and spare you the judgments of our called scientific opponents. In truth, it is almost always about people whose only certificate of suitability is the neutrality that they have accredited by staying away from the experiences of psychoanalysis.
>
> (Freud, 2006e, p. 129)

As we read, Freud assumes that he owes explanations to scientists, but stops because he was not going to respond to criticism from scientists who, for him, are not authorized. It is useless to speak to those who do not want to know, to those who tear elements from psychoanalysis, to those who are called opponents. And he expressly says that he is not going to count them among the psychoanalysts.

> As far as I know, psychotherapists who occasionally use analysis are not on safe analytic ground; they have not accepted the complete analysis, but have diluted it, perhaps they have "removed the poison" from it; they cannot be counted among the analysts.
>
> (pp. 141–142)

Freud was not going to authorize analysts who lose their principles. He makes it clear that capricious scientists cannot be counted among analysts because they have convinced themselves that they live up to the concepts and their unconscious. Not even Freud (2006c) was, and he demonstrated this in 1935, four years before his death, when he presented to the community of analysts his own failed act, a *calamis lapsus* (*calamis* means pen in Latin, that is, a mistake in writing). Freud does not believe that he has found the ultimate solution to the analysand that he is. With this, he teaches that an analyst, despite knowledge, continues to learn from his unconscious.

> To be an analyst is not to analyze others, but in the first place to continue analyzing oneself, to continue being an analysand. As you can see, it is a lesson in humility. The other way would be infatuation, that is, if the analyst believed that he was in order with his unconscious. We never are.
>
> (Miller, 2014a, p. 33)

The practitioner of psychoanalysis relies on the fundamental mutation produced in his own experience of analysis. Of course, the analyst's training is multicausal: personal analysis, study of the episteme, control of his own practice, relationship with the school. But, in order to function as the semblance of the object of the analyzing subject, the practitioner has to be aware of his unique way of enjoying. Otherwise, if no one asks you for evidence of this fundamental mutation in your own analysis, you're done for.

> The psychoanalyst on a pedestal, the psychoanalyst-magician, the psychoanalyst who is not asked to prove his aptitudes because he is backed by a powerful institution that grants him its letters of credit, is over, will be over [...] It is in the final stretch.
>
> (Miller, 2004b, p. 11)

To be an analyst is to assume error as a foundation, and it is to never abandon the analyzing position. Infatuation is what leads to imposture; to impose on the therapist who ensures knowledge. That, by not relying on his analysis, the questions he formulates in the cure he directs are compromised by the closure of his own unconscious. The neuroscientists that we present will have studied the theory, they may even say that they are analyzing themselves, but the proof of not going through their ghosts is still current in their conclusions.

The problem of scientific induction

The problem is the following. Part of proposition 1, S1 (owner significant): "all of the unconscious is located in the cerebral amygdala or in the nucleus accumbens". Because two mere observers saw the unconscious, once, twice, three times, the unconscious must be in that area; then they point out that so many other cases allow us to confirm proposition 1: all of the unconscious is seen in that brain area. No counterexample has been found, so the hypothesis is highly probable. They locate in that zone the unconscious that they "saw". There may be much more of the unconscious that they could not "see" but that many more people could "see". In addition, they "saw" it in that period of time. They can start the task again tomorrow, but at some point they will have to stop.

It is about, as Goodman (1993) says, the old problem of induction, the problem of the validity of judgments about future or unknown cases. This arises because the predictions "pertain to what has not yet been observed" (p. 96). The scientist insists "that it is necessary to find some way to justify the predictions" (p. 98). He argues that for that purpose "he needs some resounding universal law of the Uniformity of Nature, and so he inquires how it is possible to justify this very universal principle" (p. 98) which "satisfies no one but its author" (p. 98).

Thus, the problem consists in defining the relationship that occurs between statement 1, "all of the unconscious", and statement 2, "is in the amygdaloid nucleus", if and only if statement 1 can be truly said to confirm statement 2 to any degree in the unconscious = amygdaloid nucleus mode. New difficulties soon appear from other directions. If there is an unconscious that is not there, it confirms the hypothesis that all things not located in that nucleus are not unconscious. The conclusion of the future from the past has this difficulty: it is not possible to verify that it has always been so. Just because it was once doesn't mean it's always like that. There is no right *that* because it has ever been like this, it has to continue to be like that and always like that. Even if we cannot provide the minimum counterexample, that does not allow us to say "everyone". You have to pay close attention before saying all.

> That's why I take them back to this point. Those who seduce them with the synthesis of psychoanalysis and biology show them that it is manifestly a lure, not only because nothing at all points in this direction, but because, until further notice, promising it is already a scam.
>
> (Lacan, 2010, p. 364)

To promise is to know oneself committed to what has not come. Freud (2011c) said that in order to be in a position to make a judgment about the future, one must know about personal issues that are not generalizable. "And it is this, in particular, that imposes on the analyst the obligation to submit himself to an in-depth analysis in order to become suitable for an unprejudiced reception of the analytic material" (Freud, 2006b, p. 205).

> [...] Whoever, as an analyst, has disregarded the precaution of his own analysis, will not only be seen punished for his inability to learn from his patients beyond a certain limit, but will also run a more serious risk, which may become a danger to others. He will easily fall into the temptation of projecting onto science, as a theory of universal validity, what in a deaf perception of himself he discerns about the properties of his own person; it will discredit the psychoanalytic method and mislead the inexperienced.
>
> (2012e, pp. 116–117)

Projecting the future onto science is veiling knowledge about itself. Statements like "the unconscious is and will always be in the amygdala" direct interpretations under this restricted mode. Instead, psychoanalysis leads to drop the mask of ideals: "Now you see [...] where psychoanalysis leads. The mask has fallen: to ignore God and the ideal [...] as we had always suspected" (Freud, 2011c, p. 36). Indeed, when the neurotic analyzes himself something he learns, he learns to say well, he discovers that he can say no, and that is a freedom that sometimes must be given.

Senses of the term unconscious

In 1912 Freud wondered about the meaning given in psychoanalysis to the term unconscious? One answer is that the unconscious is a representation that appears and instantly disappears: "A representation – or any other psychic element – can now be present in my consciousness, and a moment later disappear from it" (2012i, p. 271). We jump to 1932 and read: "[...] we call a process unconscious when we are forced to suppose that it is activated for the moment, even though for the moment we know nothing about it" (2006d, pp. 65–66). How can we sustain that Freud wanted to fix in the brain what is characterized by its transience?

Fleeting implies discontinuous, the opposite of constant. In continuity there are no stops while in discontinuity there are interruptions, there are intervals. The idea of interval is central here. The subject emerges in the crack, between one thing and another, in the middle, like a loss, which testifies to the lack of measure, to incommensurability, it does not adhere to one or the other. Freud says: "the representation has been present in our spirit also during the interval" (2012i, p. 271). For this reason, the path to the unconscious is through openings, slits, holes, gaps, and fissures that interrupt the lexical-graphic chain and that, at the same time, produce it, while, from the relationship between the minimum pair of signifiers, trace and representation, an effect of loss is produced, the subject of the unconscious.

> It is not possible to capture the abyss that exists between the functioning of the brain and the beginning of the word in a singularity that testifies, with its stammering, with its hesitation, with its abrupt decisions, that something new and irreducible to its previous conditions of possibility begins to exist.
>
> (Ritvo, 2014, p. 215)

The brain is one precondition for this discontinuity, but it cannot be grasped by the functioning of the brain. There is no doubt that it is impossible to elaborate a concept without resorting to a closing operation. But one thing is a closure that contains within itself the unknown element, an immeasurable element that cannot be deliberately cleared, and another thing is a closure that seeks to radically expel that element.

We assume the unconscious by indices: "an unconscious representation is one that we are not aware of, despite which we are willing to admit its existence on the basis of other indices and evidence" (Freud, 2012i, pp. 271–272). We know about the unconscious by its effects, by its formations (Freud, 2007c); "[...] we call unconscious a psychic process whose existence we feel precise to suppose, perhaps because we deduce it from its effects, and of which, however, we know nothing" (2006d, pp. 65–66). And supposing it implies de-objectifying it, that is, subjectifying it.

It is difficult to prove its objectification, and this idea of the unconscious "is ultimately the only torch in the darkness of depth psychology" (Freud, 2012b, p. 159). The only torch, the only clear thing that Freud has, is that the unconscious can only be assumed, not objectified. As Miller (2010a) says, the object that interests psychoanalysis is not an object constituted in objectivity, it is not at all the object of scientific discourse. So the object of psychoanalysis raises objections to the project of making it neuroscientific.

> We don't have to use neuroscience to make them say that they say the same thing as psychoanalysis or that they confirm it. Rather, it is about distinguishing the two planes, that of scientific objectivity, on the one hand, and that of the objectality[3] of psychoanalysis.
>
> (Laurent, 2005, p. 75)

Inference is the only torch because the unconscious is studied for its formations which "[...] go beyond any physiological explanation" (Freud, 2006a, p. 44). For this reason, says Freud (2011e), "psychoanalysis [...] set limits to the physiological approach" (p. 170). "There is no possible agreement between these two conceptions [...]" (p. 173).

It should be remembered that up to 1911 Freud had been developing two meanings for the unconscious: repressed and latent. The repressed supposes an unconscious part that refers to its originality and its inaccessibility. All representation deforms the essence to which it refers because it covers the repressed part to which it adheres, that is, no representation transmits the original. The repressed unconscious is thus what cannot be externalized. While the latent goes beyond the definition of unconscious as never conscious because being latent can be done at any time, but we don't know when or how:

> [...] we can distinguish two kinds of unconscious: one that easily, under conditions that often occur, is transmuted into the conscious, and another in which this transposition is difficult, is produced only through a considerable expense of labor, and even it may never happen.
>
> (Freud, 2006d, pp. 65–66)

So, what is repressed is not all of the unconscious, it is only a part. Another part is the latent, what Freud in 1915 was going to call the non-psychic, the rest of a translation operation: "these latent memories should no longer be classified as psychic, but would correspond to the remains of somatic processes of the which the psychic can sprout again" (2012b, p. 164). If something of the unconscious is not psychic, then it is not appropriate to say that it is located in some delimited instance. The unconscious is no longer defined from a topical perspective, no longer as a system, a province. It is no longer the idea that Damásio maintains (cit. Bassols, 2011a) of mapping the brain with borders that define the zone, the space of executive responsibility of the unconscious. A map fulfills its function of representing, but the unconscious is precisely a void in what is represented.

When Freud realizes that the unconscious cannot be called a system, he becomes concerned because he couldn't close his theory of locatable instances. That is why he adds a third meaning to the term unconscious given by the singular processes that compose it. Freud changed the names of the subjects of the cases not only to preserve the person, but to call them by their jouissance names or by their symptoms: "The Rat Man" or "The Wolf Man". This "is the third meaning, and the most important, that the term *unconscious* has acquired in psychoanalysis" (Freud, 2012i, p. 277).

With this third sense, the symptom is not what must be eradicated, but what must be sustained. And when it's not there, it's what it is trying to invent. "The symptom appears as an attempt at a solution, an attempt made by the subject to face real and not as an error" (Bassols, 2011a, p. 110). Freud's genius is to demonstrate that the disease is an attempt to solve a problem (Goya, 2017). Fairly it is in the symptom that many therapists want to cure, that the subject supports himself with and, some-times, loves as himself (Miller, 2018).

Neurosciences seek to normalize. That's not bad. Institutions do it. While, on its reverse, the symptom is rejected. Each one participates in the culture with his own symptom. For this reason, Freud places the unconscious on the other side of mental health, an ideal that runs through the neurosciences.

> Is an assumption of this type, that it hosts representations in [neural] cells, gen-erally admissible and correct? I do not think so. With respect to the tendency of earlier times in medicine to locate entire mental faculties, as defined by psycho-logical terminology, in certain regions of the brain, Wernicke's affirmation that it was only it is permissible to locate the simplest psychic elements, the sensory representations [...] and this without a doubt in the central termination of the peripheral nerve that received the impression. But, deep down, isn't the same basic mistake made, whether one tries to locate a complex concept, an entire psychic activity, or just a psychic element? Is it licit to take a nerve fiber, which throughout its course was merely a physiological product subjected to physi-ological modifications, immerse its end in the psychic and provide this end with a representation or a memory image? (Freud, 2012b, p. 204)

Freud doubts that it is legitimate to unite the real and the objective. Finally, he will say that any attempt to integrate both substances will fail:

> [...] all attempts to deduce [...] a localization of mental processes, all efforts to imagine the representations stored in nerve cells and the circulation of excita-tions through nerve bundles have failed utterly. The same fate would be met by a doctrine that sought [...] to situate unconscious processes in the subcortical areas of the brain [...] Our psychic topic provisionally has nothing to do with anatomy; it refers to regions of the psychic apparatus, wherever they are placed within the body, and not to anatomical locations.
> (2012b, p. 170)

Wherever they are placed within the body, says Freud. It doesn't say inside the brain. In 1932 we read: "I will also point out that the psychic topic developed here has nothing to do with brain anatomy" (2006d, p. 93). In that year he writes about the id and says:

> [...] we call it chaos, a cauldron full of bubbling excitement. We imagine that at its end it is open towards the somatic, there it welcomes within itself the drive needs that find their psychic expression in it, but we cannot say in what substrate.
> (2006d, p. 68)

We cannot say in which substratum. There is what cannot be said where, that is, there is what has no place. The id is open to the somatic where it welcomes the drive needs that find their emergence in the body, but it cannot be said where. Freud does not intend to make a topographical location of it, he does not follow the nervous materiality. In any case, Freud follows the materiality of the signifier.

> There is a certain level of autonomy, but the ego, the id and the superego are not embodied in the brain. Of course we are materialists, but we think that this little device of the superego and the id, in short, that Freudian metapsychology, does not have to be interpreted in a mechanistic way. There is a higher materialism – "materialism", said Lacan – the materialism of the word (mot) that designates the place of the Other and its articulation with the drive, with the object of jouissance, and that cannot be applied mechanically to the distribution of the areas that light up in the brain according to the wonderful positron camera.
>
> (Laurent, 2010, p. 76)

The term "*lo*"

In Spanish we have the term "*lo*". This article does not go well with any noun, neither masculine nor feminine, which is why it is called the neutral article. It is one of the many Spanish words that do not exist in English. That is to say, a word of impossible translation. The closest words are "the" or "it", but they are not equivalent.

To understand the importance of the term "*lo*" one would have to look for its opposite term, which could be "the". Bertrand Russell (1977), an influential 19th-century mathematical logician, brings to the fore the difficulty contained in the use of the term "*el*" (in Spanish), which in English grammar could be replaced by "the". This term, says Russell, is used to designate a concept-class of which there is only one member. "The word 'the', in the singular, is correctly used only in relation to a concept-class of which there is only one instance" (p. 93). It is not the single case, but the single member in the class, the example, which by itself supports the entire class, to which is reduced a whole series of attributes that define the class and the member in an equivalent way in a relation to identity.

The difficulty this concerns, says Russell, is with the notion of denoting. When speaking of concepts, one speaks from the notion of denoting, whose essence implies a certain lack of definition, since it derives from variability, which would not be feasible in the use of the term "the". With the use of the term "the" every term would be defined; there would be no opportunity to denote, to allude, to evoke, the whole is identified. It makes a certain identity exist for which the concept and its definition matter more than the object that it really denotes and what "reasoning requires is that one work with the object denoted by the definition" (p. 94). That is to say, what the reasoning asks for is not identity, but relations.

Russell differentiates identity and relationships: "identity cannot be a relationship, since when it is truly affirmed we only have one term, while a relationship requires two terms" (p. 94). One could even question the identity that ensures the use of the term "the", since "two terms cannot be absolutely identical, and one cannot be, because with respect to what is it identical?" (p. 94). It is the idea that Lacan, relying precisely on Russell, develops in *Seminar 14: Logic of the Phantom*, that the signifier cannot signify itself.

For Russell, there must be a referent and a story. That identity is questioned does not mean that referent and story have to be different things "and where identity is affirmed, this is not the case" (p. 95). It does not happen that they are different, nor does it happen that they are related, but it happens that they are identical, which, according to Cancina (2008), closes with the meaning. The term *"lo"* allows opening the signifying game since it is not as a formal truth in any structure. Hence, it is more convenient to refer, as proposed in the title of the Spanish version of this book, to *"lo"* unconscious, which in English is translated as "the".

This was already a suggestion by Freud: "We call the unconscious the unconscious that is only latent and becomes conscious with such ease and we reserve the designation 'unconscious' for the [*lo*] other" (Freud, 2011l, p. 66). *"Lo"* unconscious to the extent that it is taken to the time of the concept, to the thinkable, implies that one has renounced *"el"* unconscious as essence, substance or principle, in this way the process of conceptualization or construction of the concept has a paradoxical aspect that is the explanatory deconstruction of the unconscious as a hypothesis (Rostagnotto, cit. Balzarini, 2023b).

The turn of the 1920s

> I'm the wound and the knife! I am the slap and cheek! I am the limbs and the wheel, And the victim and the executioner!
>
> (Baudelaire, 1856, p. 118)

The years 1920 to 1923 were fatal for Freud. Anton von Freund, who had set out to advance the cause of psychoanalysis thanks to his considerable fortune and to whom Freud was very close, died of cancer in 1920. Days later, his favorite daughter, Sophie Freud, died at the age of 26 in the middle of the Spanish flu pandemic. Three years later, his little 4-year-old grandson Heinz, with whom he was especially fond and Sophie's son, died of tuberculosis. That year, 1923, is the beginning of jaw cancer (Jones, 1981).

Meanwhile Freud's place in the world was growing and with it the pressure. As the creator of psychoanalysis, he had to train analysts and guide the psychoanalytic movement. In addition, the shortage of work, one of the consequences of the war, made it difficult for his patients to pay for the sessions while Freud bore the costs of publishing his two official journals. Added to this, he had to continue with his written work, plus his jaw cancer disease and his operations made this a very difficult time (Rosales, 2017).

Freud fights against this cancer for 16 years in pain, going through 33 operations until he dies on 23 September 1939 in the City of Gardens, London. "At the time of his death, at eighty-three, he was writing his *Outline of Psychoanalysis*, and he receives patients up to a month before his death" (Jones, 1981, p. 14). The passion that characterized him, the incessant search that he undertook, working for the cause of psychoanalysis, for pleasure, even with his body taken over by what he himself had called death drive and persecuted by an unstoppable criminal ideology that followed with his family later (after her death four sisters died in concentration camps), despite all this his commitment to the cause of psychoanalysis was unceasing.

Such an effort resulted in at least one discovery that remains difficult for many scientific quarters to accept. Following Delgado (2021), the text *Beyond the Pleasure Principle* is equivalent to traversing Freud's phantom of sustaining happiness among human beings, the phantom of well-being. That text is equivalent, says Delgado, to the fall of the supposed security of any ideal. The cruelty of the unlimited slaughter of war, man's inclination toward catastrophe, famines, expose the naked instinctual life, which prevents us from continuing to maintain a primacy of the pleasure principle. Precisely, in 1930, Freud's essential contribution will be that the human being does not achieve satisfaction, but not because of an external prohibition, but because of an internal obstacle, which evokes the idea of the impossible. Discomfort in culture, Delgado points out, is an effort to account for what is done with this impossible. Therefore, the program of the pleasure principle is not realizable. Such is the Freudian discovery that is still being denied by certain scientific sectors.

The program of the pleasure principle means life in balance. In medicine, balance is central. Everything is held in pairs of opposites: agonist/antagonist, antigen/antibody, attack/defense, oxide/reduction, loss/reuptake, pleasure/displeasure, charge/discharge. All binary and complementary. He has a reason: to treat pathologies by counteracting what is in excess with its opposite. The organ, to function, needs this balance, the result of the opposition in fair measure of two elements. Hence the opposing elements are always symmetrical. But Freud does not stop there, he goes out of the definition of unconscious opposite to conscious, he leaves physical-chemical thought, and says there is no coupling, there is no relationship.

> Well then, if there were symmetry, reciprocity, perfect coupling of the two systems, if the primary and secondary processes were exactly the inverse of each other, they would merge into one and it would be enough to operate on one of them to operate simultaneously on the other. By operating on the self and the resistance, the bottom of the problem would be touched at the same time. Freud writes *Beyond the Pleasure Principle* precisely to explain that it is not possible to stop there.
>
> (Lacan, 2008a, p. 104)

In his own practice Freud verifies that despite clarifying the meaning of the symptoms, that is, despite treating the encryption with its opposite – deciphering – the patient continues to repeat, even clings to the symptoms more than to his well-being (Blanco, 2011). It is there that Freud conceives a strange response from the unconscious that resists healing. If the analyst gives hope to the patients and shows them that he is happy with the progress of the treatment, they seem dissatisfied and, as a general rule, their condition worsens. A will in them goes against the progress of the cure, and opposes improvement. "Any partial solution [...] causes them a momentary reinforcement of their suffering; they worsen in the course of treatment, instead of improving. They present the so-called negative therapeutic reaction" (Freud, 2007b, p. 50).

> [...] it is a so to speak *moral* factor, a feeling of guilt that finds its satisfaction in illness and does not want to renounce the punishment of suffering [...] Now, this feeling of guilt is mute for the patient, it does not tell him that he is guilty; he does not feel guilty, but sick. It only manifests itself in a resistance to healing, difficult to reduce.
>
> (p. 50)

In this rare reaction, the need to be sick prevails. Freud will associate it with the cruel component of the superego. Not the part of the superego that allows socialization, but the senseless part that subjects the subject to the tyranny of moral conscience (Balzarini, 2023b). However, the negative therapeutic reaction is not negative transference. The negative transference is that a patient says I'm leaving, he leaves, he doesn't come back (Chamorro, 2011). Instead, the negative therapeutic reaction can keep a patient in that device forever.

This negative therapeutic reaction occurs when the analyst tries to produce therapeutic effects, cure the symptom, restore balance while ignoring the treatment of other issues (Delgado, 2012). The negative therapeutic reaction is a sign that partial solutions have been introduced. It is as if the patient stood up and said to the analyst: "I do not owe you my cure, I do not owe you my health".

Thus, Freud finds this paradox in the superego: the more the ego submits to the fulfillment of the ideal, the more the superego becomes punishing; the more obedience the greater guilt. Satisfaction linked to punishment. Freud says that the behavior of the ego ideal decides the severity of a neurosis. Mental health depends on the way in which this superego has been formed. If the superego is not transmuted correctly, if it does not become sufficiently impersonal, the moral component opposes the ego and asserts itself by attacking it with great severity. Dall'Aglio (2020a) recognizes that neurosciences, by denying impossibility, by submitting to the ideal of knowing everything, awaken this moral aspect of the superego. The typical phrase is "with effort everything is possible". Effort as a veil of the impossible.

With these new considerations, Freud has to reorder the theory, the orientation of the cure, the conception of the final analysis and the position of the analyst (Delgado, 2021). It will no longer be a question of explaining how a cure is produced, but of its obstacles. The negative therapeutic reaction leads Freud to modify the conception of the unconscious, an issue that seems not to have been warned in the rage for healing after efforts to localize mental pain.

Beyond scientific experimentation

Olds and Milner (1954) devised a system that allowed a rat to stimulate its own brain by means of a lever connected to an electrode implanted in the forebrain. Thus, they managed to describe that the rats would continuously press the lever in exchange for receiving nothing more than a brief pulse of electrical stimulation. It turned out that a similar effect also occurred when the electrodes were implanted in the nucleus accumbens, which when stimulated causes these unstoppable seeking reactions (Olds, 1956). Rats would press the lever frequently for stimulation, would work so vigorously to deliver stimuli to the point of exhaustion and exclusion from all other activities for survival. That is, the rat would give itself so many shocks in that nuclear nervous reward zone that it would make it abandon survival activities such as eating, drinking, reproducing and sleeping. But wasn't it according to neurobiology that the animal moved based on the pleasure principle? Wasn't the animal looking to survive?

In the wake of certain surgical procedures similar stimulations were possible for some human patients. It was observed that they preferred this stimulation over any other activity. But this stimulation was not associated with an external sign of pleasure: neither a smile nor a relaxed face, nor another sign of pleasure, of tangible happiness, or a subjective expression of a pleasant sensation (Berridge and Kringelbach, 2008, p. 15). How do neurosciences explain these divergences?

In the classic experiment, previously mentioned, the experimenter leaves the electrodes in the hands of the rat, with which the rat goes to that corner where it received the shocks hoping to receive them again. The experimenter wants to see if the rat persists in this task. He puts up a series of obstacles, but the rat still arrives, he activates the lever at his will and receives the discharges in that area of his brain. The rat begins to give itself more and more shocks, reaching the point of manipulating the lever thousands of times per minute, imagine the speed, until the only thing the rat does is cause these discharges; it stops eating, stops drinking, abandons its baby, and dies (Yellati, 2021).

If the rat is governed by the search for balance in a natural environment these discharges should not be provoked until death. The reaction conditioned by the rhythmic stimulation that the thinking being introduces is not found in a fixed way in nature. There are no rats that spontaneously search in an unstoppable way for food (Skinner) or dogs that salivate at the action of a sound (Pavlov). It is the thinking being that introduces the signifier, a material cause that is not the conditioning stimulus (food, electric shock), but the signifier in the real. To say it all, the experimenter introduces the death drive (Lacan, 2008f).

Freud discovered that there is a repetition that is not pleasant. What is repeated is something that generates suffering. It is not the repetition of the best. It goes against survival and that is why Freud will end up identifying repetition and the death drive (Blanco, 2011). The compulsion to repeat the traumatic experience opens a field that escapes repression and demonstrates, against all well-being arguments, that what is repeated is what produces pain. However, neurosciences support the principle of pleasure prevailing in the human species.

According to the pleasure principle, we feel displeasure when the tension in the nervous apparatus rises and pleasure when we discharge it. Under this principle, the rat would have to learn that when faced with a stimulus that has produced an increase in the level of arousal, it should not return, but it does return. Why does come back if it is an animal governed by the search for balance?

Neurology says that human beings seek balance, why do we continue to eat when we are full? Why does the baby keep sucking the breast even with his hunger satisfied? Why do we keep working while tired? Why don't we sleep when we want? Do we have failures in the brain centers of satiety or rest? The logic of these questions reveals that the rules of neurobiology do not coincide with those of experience.

There is an area in the brain made up of various brain structures that neurosciences have called the reward area. When someone is experiencing a pleasant situation, that area is activated, and there is a release of dopamine that occupies the receptors in that particular area of the brain. The rat has that area, and we humans also have it. Conclusion: if you want to experience pleasure, you have to learn to stimulate that area.

However, what one has to live for this dopaminergic release to occur in humans is a subjective matter. As Yellati (2018) says, for someone it can be pleasurable to watch a pornographic film, for another to look at a landscape, have a practice considered perverse and thus the list would be endless. If we follow the neurosciences we have to affirm that the act of the breastfeeding mother and the behavior of the serial rapist could show neural activity in the same region of the brain. It's scientific!

Lacan's (2008f) idea is that the rat responds to what has been cogitated by the experimenter. As we said, the rat would have to react (before the food lever, light beam, electrode), because it learned how it must behave to achieve balance in a given set-up. The rat is then defined by the body, by the brain. Lacan (2009g) recalls the passage of Henri Ey's doctrine of organodynamism, in which mental disturbance, whether functional or anatomical-lesional, is always related to the play of apparatuses constituted in the interior extension of the body's integument, to the neurobiological doctrine where mental disturbances must be related to neuronal convolutions. The rat's disturbance has to be related to some neuronal zone, so the rat goes from being the body, to being the brain organ that reacts.

What is interesting is that the experimenter manipulates instruments not to explore molecular interactions, but to make complex brain mechanisms react. For that, he has created a maze. It is the worst labyrinth because it is not that intricate shape that can trap us, but rather it is a straight, unique, and precise line in which being is the organ, which transforms the question of knowing into that of learning. The experimenter notes that the properties of the system have changed in the rat, so he can say that there was learning. A stable behavior modification is already defined as learning. It is an update of behaviorism by neurobiology, which Freud already lamented.

> Psychoanalysis [...] is enormously popular among laymen [...] Unfortunately, they have also watered it down a lot. Numerous abuses, which have nothing to do with him, are covered with his name, and opportunities to gain basic training in his technique and theory are lacking. In addition, in the United States it collides with behaviorism, which in its naivety boasts of having completely removed the psychological problem.
>
> (Freud, 2006a, p. 49)

Freud claims that the ambitions of this model feed the illusions of having solved the Cartesian problem. But the surprise comes when the experimenter introduces the signifier. There the rat shows a profound disturbance of what underlies the entire neurobiological conception: that everything we do is at the service of the perpetuation of the individual and the species. Said by Lacan (2008f): "scientific discourse ignores the unconscious" (p. 167).

Psychoanalysis does not need experimentation in its research methods. "It is only necessary to deprive oneself of the help that the experiment means for the investigation in the analysis" (Freud, 2006c, p. 161). In fact, to install the analytic discourse, a resonator is not needed, not even a couch. As Zack (2016) says, the enactment of the analytic discourse can be done without psychoanalytic scenery (if we agree that psychoanalytic scenery is couch and armchair). This phenomenal expression does not guarantee the implementation of the analytical discourse, which is not in a place, but in the love of transference. Ferenczi analyzed himself with Freud in the mountains of Vienna without the advances of current experimentation. "The analytic act can be done anywhere" (Lefort, 2012, p. 2), only the analyst has to know that "one is not involved there as I am" (p. 10). Lacan "[...] had that availability that made the analysis session possible in the car, at the table, in an explosion of paint, or on the couch" (Lefort, cit. Brodsky, 2015b, p. 28). So if the originality of the analytical method is given by the means of which it is deprived, how could the neurosciences, which precisely do not deprive themselves, be combined with psychoanalysis?

Logic of the unconscious in Freud's work

It is not easy to simplify what Freud meant by the discovery that he initially and for a long time named the unconscious. If we start from the existence of a truth that the subject does not want to know, that interferes with attention, as neuroscientists call it, that imposes itself on one's own will, then that existence cannot depend on organic functioning. Said by Freud (2011e):

> We have considered it necessary to keep biological points of view away in the course of psychoanalytic work, and not to use them even for heuristic purposes,

in order not to err in the impartial appreciation of the summaries of psychoanalytic facts before us.

(p. 184)

Thus, we have shown that from 1888, pre-psychoanalytic publications, until 1939, the year of his death, that is, for 52 years – the number of a fundamental letter in this discussion – Freud did not support integration. If for 52 years he worked to reveal the lack of relationship between the unconscious and the brain, why force his location there? As Bassols (2011a) says "[...] any attempt to localize the unconscious in the brain [...] is a pure negation of the Freudian discovery of the unconscious and a return to the psychological cave from which Freud himself had taken it" (p. 130).

Notes

1 The following biographical data is taken mostly from Jones's biography of Freud. As Trilling (1981) says, Freud authorized Jones in life because he had been his partner for 31 years.
2 In Spanish, the words "*mujeres*" and "*mejores*" sound similar.
3 In Spanish, the words "*objetividad*" (from objectifying) and "*objetalidad*" (from object *a*) sound similar

References

Arenas, G. (2018). *Estructura lógica de la interpretación*. Olivos: Grama.

Balzarini, M. (2023b). *Lo inconsciente en psicoanálisis. Un estudio preliminar*. México: El Diván Negro.

Barros, M. (2004). La salud de los nominalistas. Un estudio sobre las prácticas psicoterapéuticas. En *Revista Lacaniana. Las prácticas de la escucha y sus argumentos* (2). Buenos Aires: EOL.

Bassols, M. (2011a). *Tu yo no es tuyo*. Buenos Aires: Tres Haches.

Baudelaire, C. (1856). Las flores del mal. Poesía. Piezas condenadas. Accessed 18 November 2020 from: www.cjpb.org.uy/wp-content/uploads/repositorio/serviciosAlAfiliado/librosDigitales/Baudelaire-Flores-Mal.pdf

Berridge, K. y Kringelbach, M. (2008). Affective neuroscience of pleasure: reward in humans and animals. *Psychopharmacology*, 199, 457–480. doi: 10.1007/s00213-008-1099-6

Blanco, M. (2011). La salud mental a la luz de los cuatro conceptos fundamentales del psicoanálisis. In *Freudiana* (61) "Sueño". ELP de la EFP miembro de la AMP. Catalunya: Repro Disseny.

Brodsky, G. (2015b). Seminario clínico: "La dirección de la cura". In *Resonancias II. Revista de Psicoanálisis. Publicación del IOM2 Nuevo Cuyo*. Buenos Aires: Grama.

Bush, G. (2022). Discurso de Bush sobre la importancia de elecciones justas. Dallas, Texas, EStados Unidos. Accessed 27 July 2022 from: www.bbc.com/mundo/media-61509370.amp

Cancina, P. (2008). *La investigación en psicoanálisis*. Argentina: Homo sapiens.

Castanet, H. (2023). *Neurología versus psicoanálisis*. Buenos Aires: Grama Navarin.

Chamorro, J. (2011). ¡*Interpretar!* Buenos Aires: Grama.

Dall'Aglio, J. (2020a). No-Thing in common between the unconscious and the brain: on the (im)possibility of Lacanian Neuropsychoanalysys. *ResearchGate, Psychoanalysis Lacan,* 4. Accessed 15 April 2023 from: https://researchgate.net/publication/342870600

Delgado, O. (2012). Cap. 3: El superyó y la reacción terapéutica negativa. In *La aptitud de psicoanalista.* Buenos Aires: Eudeba.

Delgado, O. (2018). *Huellas freudianas en la última enseñanza de Lacan. Volumen III. La clínica de lo real en Freud.* Buenos Aires: Grama.

Delgado, O. (2021). *Leyendo a Freud desde un diván lacaniano.* Buenos Aires: Grama.

Edelman, G. y Tononi, G. (2002). *El universo de la conciencia. Cómo la materia se convierte en imaginación.* Barcelona: Crítica.

Fernández, A. (2019). Discurso en Plaza de Mayo. Accessed 27 July 2022 from: https://www.lanacion.com.ar/politica/volvimos-y-vamos-a-ser-mujeres-furcio-o-guino-de-alberto-fernandez-nid2314567/

Freud, S. [1895] (1986a). *Sigmund Freud Cartas a Wilhlelm Flies (1887–1904).* Buenos Aires: Amorrortu.

Freud, S. [1926] (2005). El valor de la vida. Entrevista a Sigmund Freud realizada por George Silvestre Viereck. Traducida del inglés al portugués por Paulo César Souza y al castellano por Miguel Ángel Arce. In *La Brújula* (28), Semanario de la Comunidad Madrileña de la ELP. Directora: Marta Davidovich.

Freud, S. [1925] (2006a). Presentación autobiográfica. In *Sigmund Freud. Obras Completas.* Tomo XX. Buenos Aires: Amorrortu.

Freud, S. [1926] (2006b). ¿Pueden los legos ejercer el análisis? In *Sigmund Freud. Obras Completas.* Tomo XX. Buenos Aires: Amorrortu.

Freud, S. [1935] (2006c). La sutileza de un acto fallido. In *Sigmund Freud. Obras Completas.* Tomo XXII. Buenos Aires: Amorrortu.

Freud, S. [1932] (2006d). 31ª conferencia. La descomposición de la personalidad psíquica. In *Sigmund Freud. Obras Completas.* Tomo XXII. Buenos Aires: Amorrortu.

Freud, S. [1932] (2006e). 34ª conferencia. Esclarecimientos, aplicaciones, orientaciones. In *Sigmund Freud. Obras Completas.* Tomo XXII. Buenos Aires: Amorrortu.

Freud, S. [1932] (2006f). 35ª conferencia. En torno de una cosmovisión. In *Sigmund Freud. Obras Completas.* Tomo XXII. Buenos Aires: Amorrortu.

Freud, S. [1923] (2007b). El yo y el ello. In *Sigmund Freud. Obras Completas.* Tomo XIX. Buenos Aires: Amorrortu.

Freud, S. [1924] (2007c). Breve informe sobre el psicoanálisis. In *Sigmund Freud. Obras Completas.* Tomo XIX. Buenos Aires: Amorrortu.

Freud, S. [1888] (2011a). Histeria. In *Sigmund Freud. Obras Completas.* Tomo I. Buenos Aires: Amorrortu.

Freud, S. [1927] (2011c). El porvenir de una ilusión. In *Sigmund Freud. Obras Completas.* Tomo XXI. Buenos Aires: Amorrortu.

Freud, S. [1913] (2011e). El interés por el psicoanálisis. In *Sigmund Freud. Obras Completas.* Tomo XIII. Buenos Aires: Amorrortu.

Freud, S. [1886] (2011f). Informe sobre mis estudios en París y Berlín. Realizados con una beca de viaje del Fondo de Jubileo de la Universidad (October 1885–March 1886). In *Sigmund Freud. Obras Completas.* Tomo I. Buenos Aires, Argentina: Amorrortu.

Freud, S. [1896] (2011g). Carta 52 [Letter 52]. En *Sigmund Freud. Obras Completas.* Tomo I. Buenos Aires: Amorrortu.

Freud, S. [1890] (2011i). Tratamiento psíquico (tratamiento del alma). In *Sigmund Freud. Obras Completas*. Tomo I. Buenos Aires: Amorrortu.

Freud, S. [1893] (2011j). Algunas consideraciones con miras a un estudio comparativo de las parálisis motrices orgánicas e histéricas. In *Sigmund Freud. Obras Completas*. Tomo I. Buenos Aires: Amorrortu.

Freud, S. [1895] (2011k). Proyecto de una psicología para neurólogos. In *Sigmund Freud. Obras Completas*. Tomo I. Buenos Aires: Amorrortu.

Freud, S. [1916] (2011l). 18ª conferencia. La fijación al trauma, lo inconciente. In *Sigmund Freud. Obras Completas*. Tomo XVI. Buenos Aires: Amorrortu.

Freud, S. [1896] (2011n). Manuscrito K. Las neurosis de defensa. In *Sigmund Freud. Obras Completas*. Tomo I. Buenos Aires: Amorrortu.

Freud, S. [1896] (2011p). Carta 46. Fragmentos de la correspondencia con Fliess. In *Sigmund Freud. Obras Completas*. Tomo I. Buenos Aires: Amorrortu.

Freud, S. [1910] (2012a). La perturbación psicógena de la visión según el psicoanálisis. In *Sigmund Freud. Obras Completas*. Tomo XI. Buenos Aires: Amorrortu.

Freud, S. [1915] (2012b). Lo inconsciente. In *Sigmund Freud. Obras Completas*. Tomo XIV Buenos Aires: Amorrortu.

Freud, S. [1901] (2012d). Psicopatología de la vida cotidiana. Cap. I: El olvido de los nombres propios. In *Sigmund Freud. Obras Completas*. Tomo VI. Buenos Aires: Amorrortu.

Freud, S. [1912] (2012e). Consejos al médico sobre el tratamiento psicoanalítico. In *Sigmund Freud. Obras Completas*. Tomo XII. Buenos Aires: Amorrortu.

Freud, S. [1911] (2012f). Sobre psicoanálisis. In *Sigmund Freud. Obras Completas*. Tomo XII. Buenos Aires: Amorrortu.

Freud, S. [1912] (2012i). Nota sobre el concepto de lo inconsciente en psicoanálisis. In *Sigmund Freud. Obras Completas*. Tomo XII. Buenos Aires: Amorrortu.

Goodman, N. (1993). *Hecho, ficción y pronóstico*. Madrid, España: Síntesis.

Goya, A. (2017). *Cinco conferencias sobre psicosis ordinaria*. Olivos, Argentina: Grama.

Han, B.-C. (2022). *Capitalismo y pulsión de muerte*. Barcelona: Herder.

Hegel, G. (1966). *Fenomenología del espíritu*. Mexico D. F.: F.C.E. España S.A. Ed.

Jones, E. (1981). *Vida y obra de Sigmund Freud*. Tomo 1. Barcelona: Anagrama.

Lacan, J. [1974] (1988a). La tercera. In *Intervenciones y textos 2*. Buenos Aires: Manantial.

Lacan, J. [1954–1955] (2008a). *El Seminario. Libro 2. El yo en la teoría de Freud y en la técnica psicoanalítica*. Buenos Aires: Paidós.

Lacan, J. [1969–1970] (2008e). *El Seminario. Libro 17. El Reverso del Psicoanálisis*. Buenos Aires: Paidós.

Lacan, J. [1972–1973] (2008f). *El Seminario. Libro 20. Aun*. Buenos Aires: Paidós.

Lacan, J. [1970–1971] (2009b). *El Seminario. Libro 18. De un discurso que no fuera del semblante*. Buenos Aires: Paidós.

Lacan, J. [1955–1956] (2009f). *El Seminario. Libro 3. Las psicosis*. Buenos Aires: Paidós.

Lacan, J. [1946] (2009g). Acerca de la causalidad psíquica. In *Escritos 1*. Buenos Aires, Argentina: Sigloveintiuno.

Lacan, J. [1957–1958] (2010). *El Seminario. Libro 5. Las formaciones del inconsciente*. Buenos Aires: Paidós.

Lacan, J. [1973] (2012f). Televisión. In *Otros escritos*. Buenos Aires: Paidós.

Lacan, J. [1964] (2013). *El Seminario. Libro 11. Los cuatro conceptos fundamentales del psicoanálisis*. Buenos Aires: Paidós.

Laurent, E. (1991). Psicoanálisis y ciencia: El vacío del sujeto y el exceso de objetos. In *Freudiana* (3). ELP de la EFP miembro de la AMP. Catalunya: Repro Disseny.

Laurent, E. (2005). *Lost in cognition. El lugar de la pérdida en la cognición*. Buenos Aires: Diva.

Laurent, E. (2006). "Principios rectores del acto analítico". In *Mediodicho Nº 31*. Córdoba: EOL Sección Córdoba.

Laurent, E. (2010). Interpretar la psicosis. In *Cuadernos del Instituto Clínico de Buenos Aires*, 13.

Laurent, E. (2020b). El nombre y la causa. Conicet y UNC. Córdoba: IIPsi Instituto de Investigaciones Psicológicas

Lefort, R. (2012). El camino sobre la cresta de la duna. In *Freudiana* (65) "Los espectros del autismo". ELP de la EFP miembro de la AMP. Catalunya: Repro Disseny.

Miller, J. (2018). Cientismo, ruina de la ciencia. In *Revista Lacaniana de Psicoanálisis* (24), Ciencia Ficción, EOL, 13. Buenos Aires, Grama, pp. 11–13.

Miller, J.-A. (1978). El hombre neuronal. J.-P. Changeux interviews J.-A. Miller, É. Laurent, J. Bergès and A. Grosrichard. *Ornicar?* (17/18).

Miller, J.-A. (1994b). Psicoterapia y psicoanálisis. In *Revista Freudiana* (10). Escuela Europea de Psicoanálisis-Catalunya.

Miller, J.-A. (2004b). Improvisación sobre *Rerum Novarum*. In *Revista Lacaniana. Las prácticas de la escucha y sus argumentos*, (2). Buenos Aires: EOL.

Miller, J.-A. (2004c). Verdad, probabilidad estadística, lo real. In *Revista Lacaniana. Las prácticas de la escucha y sus argumentos* (2). Buenos Aires: EOL.

Miller, J.-A. [1985–1986] (2010). *Extimidad*. Buenos Aires: Paidós.

Miller, J.-A. [2008–2009] (2014a). *Sutilezas analíticas*. Buenos Aires: Paidós.

Miller, J.-A. [2008] (2015a). *Todo el mundo es loco*. Buenos Aires: Paidós.

Miller, J.-A. (2016c). Sobre el discurso de la ciencia. In *Un esfuerzo de poesía*. Buenos Aires: Paidós.

Millar, J.-A. (2017a). El niño y el saber. In *Los miedos de los niños*. Buenos Aires: Paidós.

Miller, J.-A. (2018). Jacques Lacan: observaciones sobre su concepto de pasaje al acto. In Bardón C. and Montserrat P. (ed.), *Suicidio, medicamentos y orden público*. Barcelona: Gredos.

Miller, J.-A. [2011] (2021e). *El ser y el uno*. Los cursos psicoanalíticos de Jacques-Alain Miller. Inédito. Buenos Aires: Paidós.

Miller, J.-A. (2023). El nacimiento del campo freudiano. Conversación con jóvenes. Accessed 26 June 2023 from: www.youtube.com/watch?app=desktop&v=gAVcOuaUyYM

Olds, J. (1956). Pleasure centers in the brain. *Science Am.*, 105–116.

Olds, J., and Milner, P. (1954). Positive reinforcement produced by electrical stimulation of septal area and other regions of rat brain. *J. Comp. Physiol. Psychol.*, 47, 419–427. doi: 10.1037/h0058775

Pulice, G., Zelis, O., and Manson, F. (2019). *Investigación <> Psicoanálisis. Fundamentos epistémicos y metodológicos. De Sherlock Holmes, Peirce y Dupin a la experiencia freudiana*. México: El Diván Negro.

Ritvo, J. (2014). *La retórica conjetural o el nacimiento del sujeto*. Rosario: Nube Negra.

Rosales, J. (2017). *La valía de la escritura testimonial para la enseñanza psicoanalítica*. Querétaro, México: Fontamara.

Russell, B. [1903] (1977). *Los principios de la matemática*. Translated by Juan Carlos Grimberg. 3rd ed. Madrid: Espasa Calpe.

Sanchez R.-J. y Sanchez C.-J. (2004). Manual de psicoterapia cognitiva (fragmento). In *Revista Lacaniana. Las prácticas de la escucha y sus argumentos* (2). Buenos Aires: EOL.

Simonet, P. (2019). Claridad hipnótica del cerebro. In *Lacan cotidiano. Para Pipol 9*. Revista de Psicoanálisis (824). BOLC.

Sinatra, E. (2017). *Las entrevistas preliminares y la entrada en análisis*. Cuadernos del ICdeBA. Buenos Aires: Grama.

Soria, N. (2020). *La inexistencia del nombre del padre*. Buenos Aires, Argentina: del Bucle.

Teboul, D. (productor). Copans, R. and Cohen-Solal, A. (directores). (2019). *Sigmund Freud, un judío sin Dios* [cinta cinematográfica]. Francia: ARTE France and WILDart Film Production. Accessed 14 Septembre 2020 from: https://tv.festhome.com/ff/festival-internacional-de-cine-documental-fidba/861/182455

Trilling, L. (1981). Introducción. In E. Jones. *Vida y obra de Sigmund Freud*. Barcelona: Anagrama, pp. 5–19.

Wallerstein, R. (2004). Introducción a la mesa redonda sobre psicoanálisis y psicoterapia. La relación entre el psicoanálisis y la psicoterapia. Problemas actuales. In *Revista Lacaniana. Las prácticas de la escucha y sus argumentos* (2). Buenos Aires: EOL.

Yellati, N. (2017). Experimentar con humanos o el investigador en su laboratorio. Noche de la EOL. In *e-Mariposa* (10). Temas de psiquiatría y psicoanálisis. Revista del Departamento de Estudios sobre Psiquiatría y Psicoanálisis (ICF-CICBA). Buenos Aires : Grama, pp. 32–35

Yellati, N. (2018). *Lo que el psicoanálisis enseña a las neurociencias*. Buenos Aires: Grama.

Yellati, N. (2021). Lo que el psicoanálisis enseña a las neurociencias. Conferencia dictada por modalidad virtual a través de Yoica AC. Accessed 24 August 2021 from: https://youtu.be/O22TlWW9bLA

Zack, O. (2016). *Vigencia de las neurosis*. Olivos: Grama.

Chapter 3

The non-relationship of the unconscious and the brain in Lacan's reading of Freud

The neuroscientific unconscious and ours

Between the years 1953 and 1963, Miller (2013b) indicates that Lacan returns to Freud to say that what Freud builds around the unconscious transports the terms toward a symbolization. At least until the cut that *Seminar 7* introduces with *Das Ding* Lacan strives to show the symbolic nature of concepts that were treated by IPA analysts as imaginary. At this time the unconscious is made up of the structure of the word governed by laws of the signifier. This was what Freud demonstrated when he made a contribution to a field that had nothing to do with the biological sciences.

In 1964 Lacan reinvented the Freudian unconscious. The unconscious does not abandon its relationship with the chain of signifiers, but now it coincides with the body by being aligned to the movement of the drive around the erogenous parts. Reduced to a mouth, to an anus, to an eye, to an ear, to the hole of the drive, to demonstrate the alliance between the signifier and the body. In this epistemic change the concept of drive is defined. The unconscious is temporary, it opens and closes, fleetingly. This model of the pulsatile unconscious is the one that Lacan will take until the end of his teaching (Miller, 2011a; Cosenza, 2020).

It is no longer a question of what is missing, of what one does not have, of the knowledge that one does not have, of the memory that one does not have that gives rise to rehabilitation or compensation, but of what is not yet or is potentially (Miller, 2021d). What is potentially does not mean that it does not exist, it means that it includes emptiness. An emptiness that does not imply non-existence, but includes not-everything. With the hole, substitution is useless. It is about holding the hole to make different constructions. As such, the unconscious does not admit a harmonious relationship of the "term to term" kind because it does not remain on the side of the effects of the signifier, but the unconscious of the neurosciences is reduced to a ligature. They definitely do not admit the evasive.

It is worth mentioning that there are positions in Lacanian neuropsychoanalysis itself that recognize the impossibility of combination. For example, Blass and Carmeli (2007) and Redmond himself (2015) who argue that although the Freudian concept of the unconscious can be demonstrated on neuroscientific grounds, the same is not the case with the Lacanian concept of the subject of the unconscious, since Lacan's idea of a divided subject, characterized by a void, opposes

DOI: 10.4324/9781003458470-4

a substantial form or essential identity, hence results in a Lacanian neuropsychoanalysis being untenable.

The contemporary doctor

In 1966 Lacan says that the contemporary doctor is summoned to respond to the demand for mental health, a demand that has its roots in the right to health and that is already motivated by a world organization. If health is made public, it will be a matter of knowing to what extent it is productive. Lacan wonders: "What can the doctor oppose to the imperatives that will make him the employee of that universal enterprise of productivity?" (1985, p. 99).

What Lacan warns is that when the patient goes to consult the doctor, he does not so much ask for a cure, but rather, by placing the doctor before the test of removing him from his condition as a patient, it implies that he remains tied to that condition, he comes to ask that he be treated like the sick person who choses to stand in his illness. "Sometimes he comes to demand that we authenticate him as sick" (p. 91).

Lacan also warns that the body is made to enjoy. The doctor, if he wants to be guided by psychoanalysis, has to take into account that what is presented at the level of the body is the complete opposite of pleasure. From the death drive formulated by Freud, Lacan formulates the concept of jouissance to capture something of what Freud designated as suffering and satisfaction. Jouissance is opposite to pleasure (Laurent, 2020b). If the doctor rejects this natural dimension, his operation based on photographs will be annihilated by such an ignored nature.

> For what is excluded from the epistemo-somatic relationship is precisely what the body will propose to medicine in its purified register. What is presented in this way is presented as poor at the party where the body recently shone with the possibility of being fully photographed, x-rayed, calibrated, diagrammed and possible to condition, given the truly extraordinary resources it holds, but perhaps also that poor brings him an opportunity that comes to him from afar, namely, from exile to the one who banished to the body the Cartesian dichotomy of thought and extension, which completely eliminates from his apprehension everything that touches, not the body that he imagines, but the body true in its nature.
>
> (Lacan, 1985, p. 92)

We know that the world is moving toward the statistical model. By then the cerebral theory of the unconscious is the platform on which the current mode of this integrated doctor is based, summoned to respond to the demand for global health reduced to numbers. The power of the doctor depends on technical collaborators to take a position in the face of the growing anguish that comes from feeling called to respond to such an imperative. If psychoanalysis were included in these neurosciences, it should speak of life as organic survival, while when Lacan speaks of life he speaks of something that is enjoyed, of which nothing is known: "we do not know what it is to be alive except for this, that the body is something that is enjoyed" (Lacan, 2008f, p. 32). That is, we know about life because there is jouissance. Therefore, biological and statistical concepts are totally inadequate in psychoanalysis.

In this way, it is not about staying in the brain and what it makes possible as a distinctive criterion of superiority. It is not in the Levistraussian cut between nature and culture where we make our bet. In fact, Lacan does not oppose animal and human being, but instead opts to maintain something deeply animal in the subject. So, the opposition is not between nature and culture, but between the living and the speaking being (Bassols, Laurent, and Berenguer, 2006).

Brain, therefore I am

These neurosciences have practiced a return to the ontology inaugurated by Descartes: subject reduced to existence and existence to the brain. Thinking has become an inseparable attribute of existing. Existence remains on the plane of the verifiable by having an identical term in thought for which the mental must be located somewhere. His motto could be: "brain, therefore I am".

Indeed, the genesis of mental disorder is sought in space, more precisely in extension. They are based on the resource they find in the evidence of physical reality "fundamentally structured as the Cartesian extension [...] an extension [...] without hiding places" (Miller, 2015a, p. 168).

The symptom has to be observed, classified, and named. Injuries and deficits are displayed. But what do these injuries and deficits show us about someone's insanity? "The images of brain activation are studied, but they say nothing by themselves. It is the theories and scientists who interpret the data" (García de Frutos, 2012, p. 5). On the other hand, for psychoanalysis the symptom only exists if it is said by the subject, it originates in his own discourse (Miller, 1998d).

> Man does not think with his soul, as the Philosopher imagines. He thinks because a structure, that of language – the word [mot] carries it –, because a structure cuts out his body, and without having anything to do with anatomy.
> (Lacan, 2012f, p. 538)

Thus, Lacan highlights the exteriority of the signifier. Language comes to the body from the outside and produces a cut. It is a structure that cuts out the body and produces effects on thought. Being and thought then cannot copulate (Lacan, 2013). Instead of I am by my brain, Lacan proposes I am because I do not think: "Where I think, I do not recognize myself, I am not, it is the unconscious. Where I am, it is too clear that I am lost" (2008e, p. 108).

Body that enjoys

Lacan substitutes the term unconscious for *parlêtre* because the real cannot be touched by the structure of the signifier (Miller, 2014c; Laurent, 2016a, 2016b). The unconscious word is then imperfect because it is marked by an excessive relationship with the symbol. It is impossible to represent the existence of a real, in Freudian terms a libidinal quantum. It is evident in experience. When someone speaks, some-times they are anguished, cry, laugh, stop, the story sometimes breaks, shakes, throbs, trembles, is ashamed. This presence proves that a subject has a body, the encounter of the words with the body. The moments in which this encounter surprises the sub-ject show that the body is not one with the being. So, "it is different to say that the subject speaks with his body than to say that the organs speak in the subject" (Bas-sols, 2011a, p. 91). Thus, Lacan endows the subject of the unconscious with a body: "[…] I seat the unconscious there, with our own body" (2006a, p. 120; Miller, 2016).

On the other hand, neurosciences advance the tendency to unify organism and body. Today the human being is a body "since we have pushed man's identification with his knowledge extremely far" (Miller, 2002, p. 17). Biology is the philosophi-cal form of this imaginary. But the body reveals something irreconcilable. We are all foreigners of our own body, we all carry in our bodies something that is not comfortable, a point of disruption. There is in each subject a rejection of the body. The treatment of this rejection is, as Freud said, the singular.

"The parlêtre adores his body because he believes he has it. In reality, he does not have it, but his body is his only consistency – mental consistency, of course, because his body every so often gets up camp" (Lacan, 2006a, p. 64). The body as mental consistency is imagined as a place that lacks nothing that comes from a uni-fying idea that does not belong to the body, but to the mind. As consistency it can only be mental and the mental is always weak to weld the real (Laurent, 2016a). That is why mental health is an ideal to which psychoanalysis does not succumb. Nor is it certain that all disciplines understand the same thing by mental health. Freud said something very simple about mental health, defining it as love and work without too many difficulties.

Lacan never stopped assuming Freud as his main reference. But his last time bears the stamp of dissolution, mainly of Freudian concepts that make them go through the shredder and separates, unties itself from its reference, Freud (Miller, 2013b). That means that it no longer dissolves, it no longer fuses with Freud, it dis-solves. He dissolves his School because the theory was being repeated. He asks the followers of the Freudian Cause to mourn for the School (Lacan, 1980). Mourning is work, as read in Freud. It implies a detachment, unschooling.[1] The stickiness is related to fixation. What is fixed coagulates in a permanence from which it is dif-ficult to get out. Neuroscientists support the fixation of experience in the brain, but this text is based on the principles of psychoanalysis to move the fixation.

Note

1 Homophony in French between *école* (school) and *colle* (glue)

References

Bassols, M. (2011a). *Tu yo no es tuyo*. Buenos Aires: Tres Haches.
Bassols, M., Laurent, E., and Berenguer, E. (2006). Lost in cognition. In *Freudiana* (46). ELP de la EFP miembro de la AMP. Catalunya: Repro Disseny.

Blass, R. and Carmeli, Z. (2007). The case against neuropsychoanalysis: On fallacies underlying psychoanalysis latest scientific trend and its negative impact on psychoanalytic discourse. *International Journal of Psychoanalysis*, 88, 19–40.

Cosenza, D. (2020). El exceso en el cuerpo del hablanteser. Declinaciones y derivas en la clínica contemporánea. Conferencia. Accessed 16 September 2020 from: www.eol.org.ar/agenda/evento_escuela.asp?Evento=976/Conferencia-de-Domenico-Consenza

García de Frutos, H. (2012). Neurocientificismo, logicismo y psicoanálisis: algunos apuntes para una perspectiva crítica. In *Freudiana* (65) "Los espectros del autismo". ELP de la EFP miembro de la AMP. Catalunya: Repro Disseny.

Lacan, J. [1980] (1980). El señor A. Lección 18 de marzo. Accessed from: http://eolcba.com.ar/wp-content/uploads/2017/06/c-El-Sr.-A.-J.-Lacan-1980-.pdf

Lacan, J. [1966] (1985). Psicoanálisis y medicina. In *Intervenciones y textos*. Buenos Aires: Manantial.

Lacan, J. [1975–1976] (2006a). *El Seminario. Libro 23. El sinthome*. Buenos Aires: Paidós.

Lacan, J. [1969–1970] (2008e). *El Seminario. Libro 17. El Reverso del Psicoanálisis*. Buenos Aires: Paidós.

Lacan, J. [1972–1973] (2008f). *El Seminario. Libro 20. Aun*. Buenos Aires: Paidós.

Lacan, J. [1973] (2012f). Televisión. In *Otros escritos*. Buenos Aires: Paidós.

Lacan, J. [1964] (2013). *El Seminario. Libro 11. Los cuatro conceptos fundamentales del psicoanálisis*. Buenos Aires: Paidós.

Laurent, E. (2016a). *El reverso de la biopolítica*. Buenos Aires: Grama.

Laurent, E. (2016b). El cuerpo hablante: El inconsciente y las marcas de nuestras experiencias de goce. Entrevista por Marcus André Vieira. In *Lacan cotidiano* (576). Accessed from: www.eol.org.ar/biblioteca/lacancotidiano/LC-cero-576.pdf

Laurent, E. (2020b). El nombre y la causa. Conicet y UNC. Córdoba: IIPsi Instituto de Investigaciones Psicológicas

Miller, J.-A. [1981] (1998d). Psicoanálisis y psiquiatría. In *Elucidación de Lacan. Charlas brasileñas*. Buenos Aires: Paidós.

Miller, J.-A. (2002) *Biología lacaniana y acontecimiento del cuerpo*. Buenos Aires: Colección Diva.

Miller, J.-A. [1998–1999] (2011a). Paradigmas del goce. In *La experiencia de lo real en la cura psicoanalítica*. Buenos Aires: Paidós.

Miller, J.-A. [2000–2001] (2013b). *El lugar y el lazo*. Buenos Aires: Paidós.

Miller, J.-A. (2014c). El inconsciente y el cuerpo hablante. Conferencia pronunciada por Jacques-Alain Miller en la clausura del IX Congreso de la Asociación mundial de psicoanálisis (AMP) presentando el tema del X Congreso en Río de Janeiro. In *Revista Lacaniana* (17). Buenos Aires: Grama.

Miller, J.-A. [2008] (2015a). *Todo el mundo es loco*. Buenos Aires: Paidós.

Miller, J.-A. (2016). Habeas corpus. Intervención pronunciada en la clausura del X congreso de la Asociación Mundial de Psicoanálisis, "El cuerpo hablante. Sobre el inconsciente en el siglo XXI", Río de Janeiro, 25–28 de abril de 2016. En esta secuencia titulada "De Río a Barcelona" intervinieron también Miquel Bassols y Guy Briole.

Miller, J.-A. [1984–1985] (2021d). *1, 2, 3, 4*. Buenos Aires: Paidós.

Redmond, J. (2015). Debating the subject: Is there a Lacanian neuropsychoanalysis? *Psychoanalysis Lacan*, 1. Accessed 9 May 2023 from: https://lacancircle.com.au/wp-content/uploads/2020/09/Debating_the_subject.pdf

Chapter 4

Consequences of the biologization of the unconscious

From a paradox within these neurosciences

Descartes thought that there is *res extensa* and *res cogitans*, body and thought, two substances, no more, establishing a dualism. Neurosciences move us on from these two substances,

> [...] either for trying to reduce one to the other – which is the doomed enterprise of current scientism. Or to support their relationship, a correlation between the so-called "psychic activity" and the so-called "neural correlate", a correlation that is never fully unraveled for the simple reason that it does not exist.
>
> (Bassols, 2016b, p. 54)

What Freud says is that the extended substance is an image, the first image of the body, which comes from the synthesis that the Self makes, from fragments, of auto-eroticism, where there is still no unified image of the Self. Therefore, this extended substance is nothing extensive and is nothing unified (Bassols, 2016c).

The neuroscientific approach says that mind and body are united, not divided, ignoring the subjective division. Human life becomes an undifferentiated unit, individuality. Kandel says, "It is very likely that, throughout your professional career, neuroimaging will be able to resolve these differences in our brain. We will then have, for the first time, a biological foundation for the individuality of our mental life" (2009a, p. 389). A magnificent effort to make Freud say that "body and mind form a single entity" (Rodriguez, 2013, p. 5). We wonder how long it will take before the time comes when one walks around and asks: How much does that brain cost? "This is for you! And then you go and have it placed" (Miller, 2004c, p. 161).

We have been promised that we can control our bodies as long as we know how to take care of the brain part. The idea of integrating the increasingly complex and numerous pieces of the body has in fact already reached the tonsillar nuclei. On one hand, it is broken down into super-small units, and hyper-specific processes are isolated until reaching an atomism that juxtaposes them, and recombines them, while with the other this atomism is corrected with notions of integration and coordination. "They give us a spectral double where everything that they only present

DOI: 10.4324/9781003458470-5

to us would be integrated into disjointed pieces" (Miller, 2015a, p. 169). This "is the paradox that the speaking body, in Lacan's sense, currently accounts for" (Laurent, 2016a, p. 11).

This paradox arises from the fact that the neuropsychoanalytic approach cannot accurately delimit where this or that mental function resides, or how far the brain itself extends as an organ beyond the nervous system "when it cannot be delimited so easily in the own body" (Bassols, 2016c, p. 56). First, they try to separate the brain, they strive to make the brain the linchpin, and then they have to go back to their integration thesis. In short, the brain is an exile from its own thesis.

In the cogito, being arises through knowledge, but to arrive at unity it must be said that the unconscious arises through the organ, that is, through *res extensa*. They look for the place of unity in different regions of the brain and there the mind–body problem takes the form of a mind–brain problem (Moraga, 2019). If both substances are complementary, the singularity is superimposed by a binomial that takes the relationship between two elements and the question cannot be other than the result of a reciprocity (Bassols, 2019).

Lacan tries to stop this binomial model focused on causalism that reduces the body to matter and ignores the jouissance substance that the living body needs (Miller, 2011c). The enjoying substance is a logical modification of the extended substance that reintroduces the body at the level of the word, taking it to the level in which Descartes discards it. So, the dualism is no longer between the body and the psyche, but between the psyche and language (Bassols, 2011a). Psychoanalysis holds that there is no synthesis function while the parcel subject is persecuted by science in the form of cerebral causality (Teixidó, 2019a). What psychoanalysis calls the subject is the parcel of that non-integration (Ubieto, 2019c).

The paradox is located in the term "neuropsychoanalysis" that condenses the synthesis contrary to the part of analysis that still contains it in its name. Neuro is prioritized and left for the last analysis when its origin is said to be precisely in Freud. Neuro first, prevails, goes ahead, and sets a mode of action, psychoanalysis is its escort. "The question is to know what the science – to which psychoanalysis, as in Freud's time, can only escort – reaches what concerns the real" (Lacan, 2012a, p. 44). The real became neural (Miller, cit. Ubieto, 2019b; Miller, 2015a; Vilá, 2019a). It is confused in the scientific as it can be "explained by means of a neuronal support mechanism" (Bassols, 2011a, p. 84). In medicine, *lysis* means rupture, division, but here integration is proposed. This is the paradox.

Neuroreligion

For religion, Bassols (2011b) points out that the subject must act according to God, because God is the good. Acting well implies knowing what should be done. If you can know what to do, it is because you know what to avoid, that is, you are aware of what could happen. The subject must be able to avoid any accident, or any trauma. However, Freud discovered that in each subject the traumatic is what could not be avoided.

The thesis neuro has a God. It is enough to consult the website of *Le Point*, a French magazine, which on 2 June 2021 published the summary of the debate that arose from the activity entitled "Faced with the crisis, a fertile brain", which brought together 40 neuroscientists within the framework of the forum Neuroplaneta:

They have brought to light the traps that the crisis has set for our brain, they have identified the resources that it was able to demonstrate, and they have shared their secrets to make it our ally, in the calm and in the storm.

(cit. Castanet, 2023, p. 29)

Praise the brain. The certainty that God says what is right. If someone does not comply with that order, they have a deficit. We went from the priest who dealt with guilty subjects to the scientist who deals with deficient and disobedient subjects. These relationships between God and the brain are sealed by a dubious Spanish translation of a book by Antonio Damásio.

It is a book whose title has been translated in a somewhat curious way. In English the title reads like this: Self Comes to Mind. The self – an eminently Anglo-Saxon notion of the I – comes to the mind [...] or, well, how the I, the identity of the subject, comes to be produced as a mental phenomenon. The translation of the title that has been given in Castilian – we do not know to what extent well known by Damasio himself – is the following: "And the brain created man". It is really putting things in their place, it is a commitment to translate Damasio's neurological hypothesis into a statement with a very religious aspect, putting the brain in the place that God had [...]

(Bassols, 2011a, p. 94–95)

These neurosciences carry out an operation of conveying the order in a religious sense. There is a God who creates man and for that is where the guarantee of identification passes. The subject is the object, which hinders the work of nomination. It is "the idea that the brain would then be the ultimate cause of the human being, until it was causa sui [...]" (Bassols, 2011a, p. 95). Neurosciences are the father that regulates all the jouissance, the father who "has a brain". Indeed, they describe to us autonomous parts of our brain that do everything for us. We should be grateful to know the space for internal deliberation, the place of the internal forum.

There is at least one difference. The neurosciences seek that something can be seen so that it can be credible. Different from religion that does not need to show and the same thing has believers who suppose symbolism to the entity and it does them good. The assumption of knowing in the neuroscientist goes with the images, whereas the priest is supposed to know without guarantees. The analyst doesn't show either, so it couldn't be neuroscience.

Neither is it a religion. The psychoanalysis goes beyond the father and that is why it is not a religion. The analysand is not going to confess, but rather he is going to the analysis to say anything (Miller, 2014a), his ideas, those that come to him, but he is not going to cleanse his sins, he is not going to get rid of his guilt. Going to confess implies that something is known prior to the encounter with the Other, but going to say anything implies that one does not know something and that precisely what is not known is what is going to be discovered in the encounter with the Other. The confessional gives religion inexhaustible power because it does not deal with the real. "That is what religion was designed for, to cure men, that is, so that they do not realize what is wrong" (Lacan, 2006b, p. 86).

Imaginization of the unconscious

Currently, the international neuroscientific community is favoring, to establish its work, the plane of the imaginary, not separated from the symbolic. It is the symbolic confused by the imaginary thanks to the advancement of neuroimaging. The technological development of brain resonance allows this work to be endowed with a powerful imaginary of the symbolic. McCabe and Castel (2008) found that simply having brain images in research articles increased scientific reasoning scores compared to articles without brain images. This attests to the appeal of the imaginary in contemporary neuroscience research. As Laurent (2005) indicates, this movement is tearing the concept of unconscious from psychoanalysis to transport it, capriciously, through a biologization under which scientists from all over the world are privileging the term neuro. Such a movement reclaims its bases in the post-Freudian interpretation of the IPA, whose theorization of the unconscious was based on the imaginary axis. That theorizing is being updated in terms of neurology, but now with a new tool: neuroimaging.

The danger of this is twofold. The great importance that neuroscience has placed on the imaginary blocks us from knowing what constitutes a symptom for each subject because the relationship between image and apparatus ignores the word. The image body is paired with the machine body. Surrounded by sophisticated state-of-the-art equipment, the body becomes the object of the machine, the object of the scientific instrument, essential for illusion to exist. Such a deception leads us, by scanning the body, to the regulation of jouissance, to the regulation of the soul (Lacan, 2008f).

Let's see what Changeux, a renowned French neuroscientist, affirms: "The lightning-fast development of brain imaging methods has made it possible to identify the neural bases of our psyche" (cit. Miller, 2015a, p. 180). It is also affirmed by the neuroscientist Forest (cit. Laurent, 2016a): "Science, with its intimidating tools, provides meat for the skeleton of our spontaneous ontology [...] What better proof of the existence of a phenomenon than a supposedly faithful image of the same?" (p. 15). They are not sure something exists until it has been seen. And if the world were blind would the unconscious not exist? Suppose we are in a room and suddenly the power goes out, we can't see each other, then we don't exist? The moon in the day is not seen, so it is fake? If an object is not in the focus of attention, if it is not perceived by the senses, it does not exist. This is objectification, the illusion that there is no loss, the illusion of a cause without residue.

Intelligibility condition and cause are not interchangeable concepts. Precisely, we would be reproducing the arbitrariness and epistemological nonsense of physiology and neurosciences, which conceive the brain from language, to later make the product independent of its production conditions and make the "real" brain the cause of the symbolic.

(Ritvo, 2014, p. 101)

If cognition and cause were interchangeable concepts, pain would have to be the same for everyone. It is Putnam's thesis of multiple realizability "according to which the same mental state can correspond to several different physical states [...]" (Miller, 2015a, p. 192). That is, different brains can be reduced to a certain mental state. From the brain of a bird, a snake, a man, or an insect, we have to suppose the same sense of pain. The homogeneity of subjective pain means that what makes an insect hurt is the same as what makes a snake or a man hurt, regardless of whether their brains have different configurations. Would be possible to understand the pain of others from your own and take the generalization to a larger scale that flatten, that level, the subjective experiences. An attempt to subsume the word and its not all phallic meaning to a limited material that supposes a body that can show everything about jouissance.

Lacan's protest was precisely that the IPA analysts normalized the subject's desire in relation to an idea of genital evolution. Today, instead of starting from a genital theory, one starts from a cerebral theory. The IPA analysts used interpretations to make the subject enter a myth described in stages. Today neuropsychoanalysts look for bundles of neurons lighting up in different places in which to fit the singularity. Thus, the imagination is a defense that closes. It is very difficult to enter. It does not make fiction. "The self to which the subject of science arrives is fundamentally this tendency to close, to this imaginary coverage of the subject of the unconscious" (Cancina, 2008, p. 36). It is the tendency toward the "good form", the gestalt, the pattern, the figure, or unified structure, which is the imaginary quota that exists in all scientific production. On the other hand, the subject of the signifier, which Lacan clears, is the subject with no other reality than that of being nothing more than a cut.

Cosenza and Puig (2019) point out that Lacan's thesis of the unconscious structured as a language introduced a break in this process of imaginization the unconscious while neuropsychoanalysis updates the contemporary version of the Psychology of the Self to install a concrete neurobiologization of the unconscious, that is a true mirror of the realignment that is operating, using Freud's terms, psychiatry with neurology (Bassols, 2011a; Laurent, 2005; Bassols, Laurent, and Berenguer, 2006; Ubieto, 2019c). Kandel (2009a) credits the IPA for having made an effort to train research psychoanalysts. And the fact that Damásio was also so applauded by IPA analysts indicates the spread of this theory. In what is disseminated there is the diffuse, the dispersed, the confused, etc.

The transference is made of an emptying of knowledge. In order for the subject to be able to produce his first truths, the analyst must be deprived of knowledge.

While the realism of the image prevents such a position. First, this or that trait of thought is accentuated, by some emotion, psychic activity, for example, the dream. Second, a deterministic explanation is found for it, based on the certainty that there is a tiny specialized cerebral locality to identify what was accentuated, a locality that we will end up seeing, in the third stage, of cerebral imaging. "No discipline gives a complete representation of the mental apparatus, but a dialogue [between neuroscience and psychoanalysis] paints a more complete picture" (Dall'Aglio, 2021, p. 128). With the fixed use of an image, movements are prevented, but they speak of plasticity!

When the truth passes into the image, the real evaporates (Moraga, 2019). That is why in psychoanalysis it is not about seeing. Freud was not interested in the dream seen in the MRI, but in the account of the dream (Brodsky, 2015a). The resonance that interested him, as Lacan (2009d, cit. Bassols, 2011a) indicates, was semantic. Freud wanted the narrative of history, which is not properly history, it is not the past, and it is not a question of going to the past, as is known to be said of psychoanalysis, the chain does not go from S2 to S1, but the narrative is to tell a story, is to go from S1 to S2, it is history as a reference on which to chain the knowledge. It is to put first the subject who narrates before the subject shown. In this sense, taking Victorri's poetic approach (cit. Laurent, 2005), the psychoanalytic proposal, in response to biological theory, is to change *Homo sapiens* for *Homo narrans*. The speaking being is an animal that lies, that tells a story, its own. A spoken story. The relevance of the narrative is important even at the level of the sexual relationship. Viñal (cit. Chamorro, 2011) recalls a phrase by Javier Marías that says the following:

> [...] marriage is a narrative institution, because if it is sustained it is based on the talks that take place in bed before going to sleep, that marriage is not sustained for anything other than for being a discursive institution. These conversations make all the intimacy of the marriage, then we see how it works, but if the talk works, the marriage will work.
>
> (p. 76)

Freud (2012b) argued that postulating the existence of the unconscious is necessary and legitimate. The neurosciences also want to show that this existence is scientific, which would modify the training in psychoanalysis that should be reabsorbed by the university; the practice would also have to be modified and should be opened up to quantitative evaluations. All of which means that the psychoanalytic institution that Freud founded would have to be modified.

Peteiro (2010) points out that locating the certainty of an image in the relationship of the subject with the body facilitates a universal authoritarianism. "They rely on their mirror image, they are subjects who are oriented by a world that sends them its own narcissistic message in an inverted way: you are one, successful, autonomous" (Belaga, 2011, p. 43). It is the worldview, indicates Laurent (2020b), which takes the subject as imaginary data, which ends the Freudian concept of the unconscious. Lacan points out that the scopic drive is, of all, "the one that most completely eludes the term of castration" (Lacan, 2013, p. 85). The image is captivating because it does not know what cannot be done. So much light can blind.

If the intention is to achieve the knowledge of science, the knowledge will be abolished. With the exposed knowledge the Other is not summoned in the place of the supposed knowing how to give the meaning that is missing in the enigma, but rather in the place of certifying the evident meaning that motorizes the elaboration of science, in such a way that it can erase all singular jouissance in the ocean of statistics. With the exposed knowledge, psychoanalysis cannot be practiced. "Psychoanalysis can only be practiced based on supposed know" (Miller, cit. De Georges, 2005, p. 56). So the analytic act is beyond demonstration.

The real in neuroscience and psychoanalysis

For the neurosciences, the real is the observable reality under three dimensions in the scanner and objectively intervened. A transparent, visible real that eliminates divergences. For psychoanalysis, the real is the place of absence, between representation and drive, where the subject of the unconscious is located; what does not stop of not being written, the impossible to bear, that Lacan named "object *a*", that is the real for psychoanalysis (Bassols, cit. Laurent, 2005). Science is the incidence of the symbolic in the real (Miller, 1998a). In other words, "the real that science knows [...] is presented in symbolic form" (Laurent, 2002, p. 155). That is why they can say that the impossible of jouissance is located entirely in the neuronal and that it is modified. But it is not real of the beyond, it is not what Freud said.

For psychoanalysis, the real is what does not change, what always returns to the same place, the repetition, from which the subject cannot escape. This is how Freud discovered it under the veil of the phantasy, as something irreversible in the subjective experience and without the possibility of a symbolic realization, without a possible image that could reproduce it in a fixed way. There is no possible photography for the real. It is what never ceases to not be written in the imaginary and in the symbolic (Bassols, 2012a, 2012c). It is the characteristic of the field of invention, still to be written, whose presentation does not stop of not being written in the neural system by many mappings, scans, or magnetic resonances. This real is what neurology cannot make enter the complex synaptic network (Bassols, 2011a, 2019).

The unforeseen, the encounter, are values that Lacan puts in relation to a real that would be proper to psychoanalysis. It is the opposite side of physical determinism, opposite of everything that the quantified science of neuropsychiatry attempted (Miller, 2015a). For this reason, when Lacan (2012c) speaks of real, he gives that word a use that allows him to say that it is not biological, but that it is commanded by the entire function of significance. So, real biological, "is exhausted entirely in the adaptive communication between organisms" (Laurent, 2002, p. 78). On the other hand, the real in psychoanalysis is nothing because it has not yet happened, it is a possible response to be written before the non-relation given by the coordinates of the unforeseen (Bassols, 2013a).

The neuropharmacological model

During the 16 years of great pain in which Freud suffered jaw cancer, he avoided taking pain relievers.

> He said that he would rather think in torment than not be able to think clearly. Only when he knew with certainty that his end was inevitable, did he ask for a sedative with the help of which he passed from sleep to death.
>
> (Jones, 1981, p. 15)

What Freud rejects is not being able to think, he does not want to be a subject without his thought. Is it not this subject, as Arendt (2003) says, lacking in reflexivity that science deals with? In this neural paradigm that bets heavily on thought by locating minute areas of intervention and control in the brain, isn't the production of a subject without his thought revealed as an unwanted side?

According to Milner's thesis (1996), the subject without attributes has no conscience, has no quality, and is an empty subject, "without belief" (Miller, 2015b, p. 440). Han (2012, 2014, 2022), and this affirms that the subject today is self-exploited, pushed to hyperproductivity, which produces "the entry of his person into the calculation" (Miller, 2004c, p. 155). Freud made this subject listen in the Victorian era. "We have not come out of that era. We are entering it more than ever" (p. 157).

The current one is "the brain doping society" (Han, 2012, p. 71). Performance must be ensured, even "some scientists argue that it is irresponsible not to use these substances" (p. 71). You just have to ensure that they are available to continue being alone. But, Handke (2006) points out, despite the pill, the fatigue persists. It is not an eloquent fatigue, which allows one to reconcile, look, and give way, but a fatigue alone, a fatigue in silence, but with the noise inside.

In the city of Córdoba, Argentina, the value of the Rivotril tablet, for example, retails, at this moment, for $100. While the value of minimum ethical fees for individual psychotherapy as recently regulated by the College of Psychologists of the Province of Córdoba is $1800. Why pay for the presence of the Other if I can enjoy it at a lower cost?

What Kandel (2009a) tells us is that the pharmacological model is at the same level as the psychotherapeutic model: the level of the synapses. The problem is that the invasion of molecular biology and medicines produces a great lack of interest in the clinic and prevents the establishment of stable relationships with the patient. "There is no more interest in thinking about clinical phenomena, but from the effects that can be obtained with drugs, which became the organizing principle of the clinic" (Miller, 1998d, p. 167).

This problem originates from the fact that the medication produces changes and this feeds a hypothesis that says that the cause of subjective discomfort is in the brain. Chemistry becomes a divinity of the real, daughter of doxa and episteme, which realized for the invasion of a deregulated and limitless jouissance. This seems to reassure many patients while some scientists can certify that the pain is a known, chemical and autonomous mechanism that does not obey any other causality; that it unfolds in an identical way in all people, that manifests itself within the stable and uniform group of those affected. The reassuring illusion that the affection is predictable, that it has an assigned, average, normal and regulated period, implies presenting oneself as a person devoid of singularity, a sample of a nosographic class, an impersonal representative of a supposedly universal knowledge that does not justify work of subjectivation, in short, reject the unconscious (De Georges, 2005). That the antidepressant animates, that the anxiolytic calms, that the stimulant improves attention, suggests that the cause of the disorder is neuronal (Briole et al., 2019). It is true that psychotropic drugs can be useful to the extent that they are administered prudently and serve to facilitate speech, but not to block it (Zack, 2007). It is also true that not all subjects are in a position to accept the unconscious; many times the chemical resource is a way of supporting the rejection of the unconscious and the analyst must be cautious before orienting from the perspective of the signifier (De Georges, 2005).

Science has advanced to provide the patient with some transitory and urgent appeasement. The pharmaceutical industries pressure the committees to establish new nosologies that feed the promise of completeness and there the drug goes from therapy to marketing. It goes to the rhythm of the present. Today we live in a deeply medicalized society, full of prescriptions. There are those who resist falling into the hands of this standardization machine, but they pay the price of being classified as misfits. For them there is an argument of sufficient moral control. Knowing the point of dysfunction, knowing the location of the subjective pain, one intervenes with the latest generation of drugs. Even love can be regulated!

> An anthropologist lady, being cognitive, wrote a work on the chemistry of romantic love. It defines what it is to be in love: it is to see your serotonin level drop to less than 40%. This was verified, measured in guinea pigs, selected [...] those who claimed to think at least four hours a day about their loved one. It has been found that they had at least 40% less serotonin than average.
>
> (Miller, 2015a, p. 135)

"If love is really correlated with a 40% decrease in serotonin, it means that it really exists" (p. 141). The brain becomes the place where everything happens. For example, at the level of the statistical manuals of mental illnesses, the DSM lists conduct disorders that are mostly treated with medication, while the significance of the symptoms disappears.

It is true that in many cases the psychoactive drug was the supplement that contributed to regulating the body, which made many screams of psychotic patients stop, which allowed many subjects to walk again for a while, that is so, the drug can accompany, until the analyst object is entered that can then be dropped.

It is from this point of view that psychoanalysis and psychiatry differ with respect to, for example, the way of approaching patients with suicide problems. While the psychiatrist is going to seek hospitalization to erase the idea of death, the psychoanalyst, indicates Chamorro (2011), is going to make the subject speak of that idea so that it can be delibidinized. If the psychiatrist works against death and on the side of life, the psychoanalyst works on the side of death, but not against life. For example, to a patient who has made a suicide attempt for the third time, the analyst says: "Talk to me about death and tell me why you failed for the third time". This is working on the side of death, but not against life (Balzarini, 2023a).

The health professional goes for life, the law obliges to go for life. The law requires health specialists to intern, to do everything to erase bad ideas from patients and for that they have psychoactive drugs, while the psychoanalyst's instrument is the word. With this instrument, Freud was able to dismantle the cathetization of bad ideas in his patients. The psychoanalytic treatment goes through giving the floor to the subject who suffers, not by crushing it with the prescription. It does not mean that the psychoanalyst shies away from medication. Medication is available for the unbearable, and analysis is available for suffering. Faced with this growing trend toward medicalization, psychoanalysis proposes the desire to know as a healer. And there the transference love is fundamental.

The right to health

Mental health has become the most precious asset. You have to be in good health. That is why the advice of neurosciences that comes from the side of withdrawing from peers is promoted: fresh air, pure air, and a "getaway" is recommended. Is mental health something that means withdrawing from the human beings that make up this world? Is harmony the search for a space in a world that is made by man, but not for man?

Today, new items are constantly offered on the market, fashions change, and when they are consumed, they are no longer new. The enthusiasm of the new always enjoys good publicity. Faced with its direct opposite, tradition, it seems the perfect way to forget the past. The truth is that talking about something truly new in a person's life is not that simple. Those who make a change do so through personal work that allows the inauguration of new ways of enjoying (Ordóñez, 2023).

In hypermodernity, the new lasts less and less time. Objects quickly become obsolete and that is where the imperative occurs: to turn toward something immediately new. Neuroscience thus goes to the foundation of this economic system: "I know what you are". Instead, the door remains open only if the subject is made to understand: "I don't know and so you must speak" (Miller, 1994b).

Nikolas Rose (2011), a post-structuralist sociologist, argues that the modern state aims to create regimes of truth based on the paradigm of biology. According to this sociologist, nothing is impossible in biology. Finding a way between the impasses of each subject is like asking these sciences to cease to exist. There is no science if it is not with capitalism. That is the scientific abyss.

Judith Miller (2018) affirmed that if science is captured by capitalism, we cannot place it among the discourses. For Lacan (2008e), science assimilates knowledge in the place of truth, promising it through scientific work, and for this reason science does not establish a social bond. Instead, psychoanalysis is a discourse. We know from Lacan that in speeches the truth is half told, that is, not the entire truth can be said when we speak. For there to be a discourse, says Miller (1994b), the analyst must show himself to be inhabited by a stronger desire than the desire to be the master. And this is still not intelligible by neurosciences.

In sciences, you have to be competitive, specialize, and tear apart the object of knowledge. There are 200 investigations that compare the numbers of subjects suffering. The comparison is "the highest form of the humanity" (Miller, 2015a, p. 138). It is the condition for commercialized results. Says Solms: "Surely it is also about the 'commercialization' of the research results and this is something that I do not intend to hide at all" (2006, p. 75). Indeed, in order to maintain themselves, experimenters must publish and for that they must choose productive lines. Research is carried out on the unconscious in the ventromedial zone of the brain, it is legitimized, advice proliferates, everyone buys, everyone arrives at the same place, all along the leading line. There is a need for meetings between professionals from rehabilitation teams to combine criteria and unify treatment objectives. All to the same side. You have to agree. There can be no divergences. The goal is adaptation.

The neuroscientist has gone from the love of knowing to the commercial desire. "We have gone from a scientific era to another marked by its intimate relationship with the market. A mercantile scientism emerges in which science serves to realize what is possible and sell it" (Peteiro, cit. Bassols, 2011a, p. 211). It is the evidence that science seeks to standardize: for all persons this unconscious. The neuroscience and market alliance is difficult to break.

Comparison is the basis of this alliance. For Pinker (2004) unconscious desire could be known while learning how others have evolved. Happiness in the human being would be something that is measured in comparison to what generates happiness for other people.

> How can a brain know if there is something worth striving for? Well, he can look around him and see how well other people are doing. If they can do something, maybe you can too. It is other people who set your scale of well-being and tell you what you can reasonably expect to achieve.
>
> (Pinker, 2004, p. 147)

So for Pinker the brain looks at others. This is opposed to psychoanalysis. As Miller (2015a) says, psychoanalysis "offers some resistance to conforming to the regime of the homogeneous" (p. 134). The patient is not even compared to himself from one session to another. The patient does not necessarily say the same thing or have the same angle, the same perspective, every session. The rare, the diverse is not tolerated.

What is rare, says Miller (2016b), is the psychoanalyst, rarer than his own patients. His interpretation is rare; it is not balanced, it is not ordered, it does not have a thread. To have a thread is to synthesize, it is to seek to tie the diversity and prevent its reinvention. The advance of these neurosciences blurs the border between knowledge and invention. This quantifying operation unfolds like a roller, invading educational, social, political, and clinical spheres. There is no reform that is not carried out in the name of this mystique of evaluation (Esqué, 2017).

Neuroscientificism

In 1951 Lacan (2009e) denounces that he is a witness to the deviation of psychoanalysis under various forms of "psychologism".

> We see them, then, under all sorts of forms ranging from pietism to the ideals of the most vulgar efficiency, going through the range of naturalist propaedeutics, taking refuge under the wing of a psychologism that, reifying the human being, would reach outrages next to which those of physical scientism would be nothing but trifles [...] in short, a homo psychologicalus whose danger I denounce.
>
> (p. 211)

We know that physics quantifies the phenomena of the universe. That is an insignificant thing if we compare it with the movement that reifies the human being using readjustments in his technique derived from science. This is what happens with this movement within psychoanalysis, which this book shows is not the most productive, wanting to turn it into a scientific method at all costs. There the substance of what Freud discovered is lost, that the unconscious cannot be an object in the current terms of science. Bassols (2013a) indicates that scientism "is not science, but one of the effects of science, which reduces everything that is subjective to something quantifiable, evaluable by numbers and observations" (p. 120).

> Scientism is extrapolation, the extension of the scientific method taken to all areas of human life. All too generally, this scientism is based on a mechanistic reductionism that assumes a mechanical relationship of a supposed cause to an effect, a relationship that would thus explain the vast array of phenomena of human life. The entire scope of the subjective would then have its causal and mechanical explanation in a determination of parts of the real isolated in the human organism. This determinism would seem to be written as the script of the book of life, a script that today is believed to be encrypted in two fundamental places: in the genes and in the neurons. These would be, according to a more or less deterministic vision of the human being, the places where the destiny of the life of each subject is written, a destiny that science would also be in a position to be able to rewrite.
>
> (2011a, p. 74)

One of the branches of this scientism is what we propose to call neuroscientism, which relies on the neuro signifier to feed faith in a subject devoid of castration (García de Frutos, 2012). It is the version of science that does not renounce the dream of positivist objectivism to understand, under the image of the scanner, that the unconscious is a product of an organization in the wiring of the brain, which can be detected, repaired and controlled; it is the mask of eminent scientists who use terms associated with psychoanalysis,

> a certain use of science that would reach all corners of the human being to manage, repair, promise a certain good under the idea that by managing our central nervous system we are going to be able to eliminate subjective discomfort.
>
> (Bassols, 2012b, p. 5)

Freud warned this movement embodied in psychiatry (Brodsky, 2013). What Freud was able to notice is that "there is no subject that is involved in the natural sciences" (Castanet, 2023, p. 74). To stop this he signed with his fist that consciousness and the nervous system existed, but he did not give proof of their relationship:

> Of what we call our psyche (psychic life), two terms are familiar to us: firstly, the bodily organ and scenery of it, the brain (nervous system) and, on the other hand, our acts of consciousness [...] Instead, we are not familiar with what is in between; we are not given a direct reference between the two terminal points of our knowledge.
>
> (Freud, 2006g, p. 143)

There is no proof of the relationship between the two points, said Freud in 1940. This does not mean that Freud is against science. Both Freud and Lacan proposed theorizing in the hands of some science. It cannot be claimed that the search for a material base is alien to psychoanalysis, on the contrary, it is present from the beginning to the end, and it runs through both Freud's work and Lacan's teaching. But all that effort on the side of materialism rejects experience (Miller, 2015a; Lacan, 2009a). Indeed, neuroscientific conclusions arise from observations rooted in laboratory investigations with animals in artificial environments, while psychoanalytic responses arise from the analysands. Hence, there is a tension between neuroscience and psychoanalysis that makes it difficult to validate unconscious demonstrations, since this proof takes place in the experience of a subject (Miller, 1987a).

Freud never believed that the achievements produced through the experimental method could modify the hypotheses derived from psychoanalysis. Lacan not only did not believe that the achievements of science could question psychoanalytic findings, but also questioned the conclusions resulting from scientific experiments (Yellati, 2018).

It is true that science and psychoanalysis share one thing: both proceed through the word. In both cases it is logotherapy, language therapy (Miller, 1994b). But under the label "psychotherapies", Miller (2013b) points out that a series of practices has been covered that goes as far as gymnastics. Gymnastics would not be the most harmful, it is even a highly recommended exercise. The problem is when some forms of receiving the demand are close to the analysis, are inspired by psychoanalysis, that is, when that gymnastics refers to the duty to perfect reflective skills.

"If we take things to the extreme, let's say that there are forms that claim to conform to psychoanalysis, and if we want to take them to the extreme of the extreme, let's say that they claim to be psychoanalysis" (pp. 101–102).

So, the problem is not formulated in terms of whether psychoanalysis is a neuroscience, but rather it is formulated in terms of a conflict of causality: is all causality that concerns the human being only of a material nature or is there also another order of causality? Do the disorders that affect the human being all respond to a neurological causality or is there a causality that constitutes an original order of reality?

It is not about repudiating scientificity, not at all. The brain is a condition of possibility, but not a condition of existence, in the same way that the physical order is a condition of life, but not its ultimate cause. Nor is the subjective cause definitive. The condition of existence that we build is necessary, but not sufficient, because the signifier cannot be made the first cause (Ritvo, 2014).

Thus, this neuroscientism is the effect of the alliance of science with capitalism. An effect of science that will always be reluctant to psychoanalysis, a discourse of crisis and not of conformism. Integrationism imposes the experience in a justified psychologism in a model like behaviorism whose mechanism is certain determinism. It is a reductionism, which raises his narcissism whose aphorism indicates that the psyche is the same as the organism. Taking Freud by parts, partisanship, will then drag psychoanalysis toward a new "ism".

Science fiction

In an interview conducted by Emilio Granzotto for *Panorama* magazine in 1974, Lacan answers the question: what is the relationship between science and psychoanalysis? He says that the science that puts the laboratory on the altar will never be able to find the truth because the truth can only be told as a lie. And that the only science he respects is science fiction:

> For me, the only real, serious science to be interested in is science fiction. The other, the official one, which has its altars in the laboratories, gropes its way forward, unbalanced. And he even begins to be afraid of his shadow. It seems that the moment of anguish has arrived for the scientists. In their laboratories […] they begin to wonder what could happen tomorrow, what their always innovative research will end up contributing. Anyway! – I say. And if it was too late? The biologists ask it themselves, sparing the physicists, the chemists. To me they are crazy.
>
> (Lacan, cit. Martinez, 2018, p. 193)

They are crazy, says Lacan, which is another way of saying that psychoanalysis could not be combined with biology. And he says, there is a serious science, the science of fiction. In effect, the subject goes to the analyst with a fiction that has been built, and goes to find out the story that he tells himself. The fiction that is built into the analysis is an interpretation that allows it to function (Pino, 2018). Making fiction is different from making memory, it is the creation of a new signifier because the referent is not denied, it is lost (Koretzky, 2019). We never recover exactly what provided us the satisfaction. There is a loss of that first cause, that perfect return is impossible. There is a "lost" (loss) that is a "lust" (pleasure), a lost that we designate unconscious (Bassols, Laurent, and Berenguer, 2006).

Thus, psychoanalysis cannot postulate first causes. "A first cause is precisely what we never apprehend and, precisely for this reason, the notion of fiction was born [...]" (Ritvo, 2014, p. 101). The neuroscientific operation has focused on identifying the neuronal cause as the first. The axiom "the unconscious is in the brain" seals the signifying and signified alliance. On the contrary, what makes it possible to operate in psychoanalysis is the disjunction between signifier and signified that makes interpretation pass for the enigma. The analytical operation consists of relaxing the ties established between the signifier and the signified, in stopping the passage from the signifier to the signified, at least slowing it down and awakening another signification (Miller, 2016b).

"The neuro thesis uses the term unconscious, but it is a corticated unconscious" (Naccache, cit. Castanet, 2023, p. 26). All altered subjects have a failure in the neocortex or in executive functions. That must be somewhere to identify. Identification must be taken in the exact sense of location. It is the opposite of creating. The opportunity to create comes when something is not in its place (Miller, 2021d, 2015a). It is a place, which is not in the brain, but in the loss. That is why fiction is not complete, "it is the consequence that it is impossible to complete" (Chamorro, 2011, p. 133).

If that impossible is rejected, there is no fiction, there is fixation. The task of the psychoanalyst, says Laurent (2020a), is to shake the fixation. It does not mean that the subject is left without fiction, without ideology, it is not that he does not have his faith, his love, his belief. You cannot go through the world without semblances, without partners, without ties, without anchors, because with the cynicism of the naked drive everything is jouissance. The semblances are a defense against reality insofar as it is impossible to bear. But that his belief does not have it in relation to ideals allows a more dignified love, more ding, with more thing.

An unworthy love finds its foundation in the belief in the existence of the sexual relationship. Getting rid of that annoying relationship would make love a little more dignified. The knowledge obtained in the transference relationship makes love more dignified. Precisely in a psychoanalysis it is about dressing the real with some fiction that allows enjoyment to be experienced. The psychoanalyst "tries to detach the subject from reality to introduce it into the rhetoric that leads to a fiction, because we want to understand the effectiveness of making fictions, which Freud called constructions in analysis" (Chamorro, 2011, p. 133).

In the 1990s, a recurring segment called Pinky and the Brain was projected in an animated series, a show created by the great director Steven Spielberg that dealt with two fictional characters, two genetically altered laboratory rats, who dreamed big. In each episode Brain, a mouse with great intelligence and grandiose desires, devises a plan to try to take over the world, and Pinky, another genetically altered mouse but less intelligent than Brain, is his secretary. This dialectic of the secretary without attributes, who lends his material for the agency of a discourse used by the ambition of the conqueror, has resurfaced, but with the difference that now it is not fiction and that is its greatest threat.

The ascent of the signifier neuro

As we have been seeing, the shape of the world today is suffering an identification to the neuroscientific model that is increasingly infatuated with dragging aspects of human life toward a dominant signifier (Miller, 2018c). Indeed, we are witnessing "a massive presence of the signifier – neuro – in our world, a prefix that has become what psychoanalysts call a master signifier, that is, a signifier that explains and serves all uses" (Bassols, 2011a, p. 84).

For Lacan (cit. Miller, 2015a), the master signifier is the unique insignia that bears the attributes of power: scepter, crown, throne and also bears what is imposed. It is what is said first, and what is said first decrees, legislates, and confers its obscure authority on the real.

> Right now we are full of the neuro prefix. Everything is neuro, neuro-justice, neuro-legality, neuro-equality, neuro-inequality. Everything is going well, we are on the right or on the left only because we have the cables made one way or the other, etc.
>
> (Laurent, 2020b, p. 74)

Neuroeconomics involves observing the electrical activity of the brain while making investment decisions (Camerer, Loewenstein, and Prelec, 2005). Also in legal issues with the lie detector theory (Langaney, 2006; Slezak, 2018) and the prediction of criminal behavior from associated brain areas (Stagnaro, 2009). Also the neuroreligion "since by observing the brain during prayer they verified how much good it does to the neurons, they did a survey and the belief in God can also be built on an image" (Miller, 2015a, p. 141). "And so, evidently, all aspects of human life are likely to be neurologized in this way, everything activates the brain [...] all human activities are likely to have neuro – in front of them" (p. 142). "There is a 'neuro' for everything and for anything" (Bassols, 2011a, p. 84).

What differentiates psychoanalysis from other therapies is the way to extract certain singular signifiers, an operation that is hampered by this domain of the neuro signifier whose prevalence is in any case. When I'm sad, my brain observes a number of events in which one has the idea that thought and neuron go hand in hand. An homologation is made between certain thoughts and certain neural networks, the correlation is sought, but not the causality. "Obviously, when you start from that, the only thing you find are mental disorders, dysfunctions" (Miller, 2004c, p. 161). Thus, "mental disorders are diseases like the others, like the flu or diabetes" (Lardjane, 2019, p. 12).

"The scientific approach implies that everything that makes us human must be measurable" (Peteiro, cit. Bassols, 2011a, p. 203). A derivation of Galileo's scientific operation that was based on measuring everything that is measurable and making what is not measurable. The novelty that neuropsychoanalysis brings is that the long-awaited statistics, the model toward which the world is moving, can descend to each subject. It is about the illusion that the problem between subject and object, the problem of the subject being rejected in scientific research, would be solved thanks to the fact that the prefix neuro has managed to introduce quantity into the one by one. Then "the description of the brain activity of a subject" will be made (Miller, 2015a, p. 145).

It is about the replacement of psychic reality by psychic activity. Psychic activity is the set of molecular interactions, neuron by neuron, in the cerebral extension that gives a harmonious continuity. But activity does not change things. The activity accompanies the development, "but sutures or forecloses everything that concerns the registration of the act" (Miller, 2015a, p. 180). What Lacan proposes (cit. Miller, 2015a) is that existence, if it is accessible, is so through the chain of words and, especially, through the one that unfolds in the analysis, which offers a landscape very different from that of the psychic activity.

Dall'Aglio (2020a) affirms that this rise of the neuro prefix has allowed many disciplines to open their doors to neuroscience, to unify with them and thus improve their efficiency. It is the figure that manages to capture the unconscious, where the being finds its guarantee. We are on the way to the progressive extension of the reduction of all human activity to mere cerebral observation. Thus, this prefix aspires to be a miraculous signifier that will allow controlling ways of enjoying (Brousse, cit. Ubieto, 2019b, 2019c). But, in clinical institutions there is talk of neurodiversity, which opens a paradox, equality, or exception.

Thus, these "neurosciences would already be demonstrating the Freudian hypotheses and, therefore, in this way, psychoanalysis would finally be a part of neurosciences" (Bassols, 2011a, p. 84). Solms is "at the head of this mix in which he claims to have shown that the unconscious [...] has its precise place in the brain, following a neuron-based functioning" (p. 84).

Psychoanalysis position

The time has come when we must ask ourselves: what is the place from where the psychoanalyst intervenes like? Is it a quiet place? It is secured, it is protected? We must say no, as it is not entirely on one or the other side of the strip (Miller, 2004c). Its place is in relation to the restless. The restless comes from the strange. For Lacan, the unconscious is a knowledge "strange to the discourse of science" (2008e, p. 95). Freud himself becomes strange to these neurosciences, they cite him to accuse him. Damásio confesses that when he reads Freud to prepare for a conference on neurosciences, he gets a mixture of admiration and irritation, a meeting of ambivalent feelings that reveal this game of attraction and rejection. He says it like this: "I spent weeks reviewing Freud's texts, oscillating between irritation and admiration as always happens to me when I read him" (cit. Bassols, 2011a, p. 129).

Lacan invented the term "extimacy", which Miller (2010) deepened as the strangely familiar, the so intimate that it becomes alien, "what is closest, most interior, without ceasing to be exterior" (p. 13). Extimacy is not opposed to intimacy, it is not its opposite "because the extim is precisely the intimate, even the most intimate" (p. 14). "What is the extim? The unconscious" (p. 20). We could say that the place of psychoanalysis is of a profound extimacy (Bassols, 2011a). This does not mean that its place is outside, but, on the contrary, between, inside, but in a space that is not scientific, at the juncture, at its exact fissure, at the interstices (Miller, 2012a).

As early as the 1970s, recalls Bassols (2011a), Lacan affirmed that psychoanalysis is a discourse of practice that "adjoins from outside with the discourse of science" (p. 26). It arises from science, but it is not science; he is the son of science, but an illegitimate son (García de Frutos, 2011); it could not be understood without it, but it raises objections of principle when it comes to human suffering, which can never be reproduced experimentally (Bassols, 2016a). In fact, to work, the psychoanalyst has to speak, a little, the language of the Other, but to tell him what he does not want to hear.

Revolution or subversion

In the 15th century, Copernicus leads a revolution, an affront to the narcissism of the human being, when he showed the world that the theory of geocentrism, founded by Aristotle, was wrong. From then on the earth is no longer at the center of the universe and with this Copernicus overcomes monocentrism.

In the mid-19th century, Darwin leads another revolution, a second narcissistic wound, when he demonstrates that the human being is not the master of animals, he is not separated from them, he even descends from them, so the human is not a divine creation and with this there is no reason to believe himself superior to animals.

At the end of the 20th century Freud produces a third narcissistic wound, he knew how to hurt the arrogance of the human being by revealing that the self is not a master even in his own house, that the human being is not sovereign of his own soul (Freud, 1986b), it hurts the extreme self-love of humanity by destituting one center for the benefit of another, the unconscious, which "in fact shows the need to reduce the arrogance that sustains any monocentrism" (Lacan, 2009c, p. 444).

When arrived in the 21st century, neurosciences changed the center again. It is about the affirmation of thought "from the study of a living organ, the brain" (Miller, 2015a, p. 144). The behavior is no longer observed, now the neurons are observed. This is the revolution of the 21st century.

As Lacan (2008f) indicates, the starting point of every revolution is what turns. The earth revolves around the sun, reproduction revolves around natural selection, the human species revolves around the signifier and now around the brain. What matters is the center. The revolution is founded around a central clot that affirms the return, which always returns to the same point. "And what remains in the center is that old routine according to which meaning always retains, in the end, the same meaning" (p. 55).

To revolutionize is to revolve repeatedly around a center. The neurosciences are looking for the locality of that which speaks. It is disturbing that more than a century ago Freud discovered something that X-rays, tomographs, and other observation devices have not yet been able to elucidate: who is speaking and where is that which is speaking located (Bassols, 2013a).

That is why Lacan does not change the center, he makes it fall, that is, he takes away the consistency of the scientific law that a center must exist, he relaxes that obligation that every matter must have to revolve around a meaning. The center is no longer where an entire humanist tradition assigned it (Lacan, 2009c). With Lacan the center is an enigma, it passes through the object, to release it. It is a lost object on which the life of a subject revolves, in an analysis the subject understands that circuit, he has to go through it, to give it up, hand it over, pay with it. Being as a substance falls, neuro-ontology falls and ethics remains. The psychoanalytic cure is based on an ethical position that does not have a single center. We now release the imaginary grasp of the meaning. However, this is difficult to accept. Today people are lost, looking for a manual on the internet, looking there for the most unlikely things. If it is convenient for the pacifier to be placed like this or by the handle, with a strap or not, if the diapers should be of that brand, in short, people now need sentimental technical assistance, that is why self-help books, blogs, proliferate from the internet, the coaches, the people who are supposed to psycho-train you. It tends toward an increasingly invalidated society. We are less and less prepared with the feeling that we have to turn to expert sources of knowledge to advise us on everything: how to breastfeed, how to buy clothes, what soap to buy. Such a state of mental weakness is reached that we become incapable of making decisions if we do not consult some kind of external referent (Dessal, 2016). This center is expected to provide everything, to be at the service of filling what I lack.

This is what Manes (2019), a neuroscientist who has acquired press in Argentina, proposes, stating that there are seven tips for taking care of your brain health that, if followed, guarantee mental health! Indeed, neurosciences know how to live a healthy life. They have become that center. It is a perfectly spherical conception, without discontinuities, a revolution. Instead, psychoanalysis refrains from inventing new imperatives (Laurent, 2014b).

Desire to wake up

In this way we have dealt in this chapter with the clinical problems resulting from the operation of identifying the subject with a signifier that can explain his relations with the other, and from there achieving a state of scientific peace, a suspicious peace that ignores the subjective matters that are not made to go to sleep, therefore the word we propose as a guide is wake up. As Stecco (2020) points out, Freud went beyond the decipherable of the dream, awakening psychoanalysis from the dream of meaning and fully interpretable. Since then, psychoanalysis cannot dedicate itself to encryption. So Freud's intention was on the side of awakening.

The desire to awaken is linked to the term youth, but it has nothing to do with age, but with vitality, with drive, with the desire placed on some cause. Youth is a logical question, not chronological. There are colleagues who have a long history who are still very awake and others who, as Miller (2005) says, are less surprised than the others because they feel like they are already experienced people. It is true that we can learn in the form of regularity, in the form of verification, but we also learn, perhaps better, in the form of surprise. "However, does not imply that in practice one wants to be surprised. Many efforts are made to become experienced, there is a lot of emphasis on that" (Miller, 2005, p. 17). Freud does not remain in the desire to sleep, he does not remain in the amazement of what is known, of elements already known, for example the amazement of climatic phenomena such as snow. Freud settles in the desire to awaken, which operates beyond repetition, he seeks to animate, cause work, which is not easy, it is not easy to settle in the desire to awaken from the point of view that it is not something that lasts, but it is discontinuous, it is eruptive.

What awakens is the opportunity to have to do something with the nonsense. And that is not possible in the dream of these neurosciences. What is expected of an analysis is to wake up. Waking up interrupts the sleeping state. The practice of psychoanalysis is a practice reduced to scansion, "it must have awakening as its goal, in the sense of the emergence of the real" (Acosta, 2020, p. 31). It is the opposite of what these neurosciences propose, a practice reduced to the continuation.

References

Acosta, S. (2020). *Die Traumdeutung*, entre el deseo de dormir y el despertar. In *Scilicet* El sueño. Su interpretación y su uso en la cura lacaniana. Publicación en razón del XII congreso de la Asociación Mundial de Psicoanálisis. Buenos Aires: Grama.

Arendt, H. (2003). *Eichmann en Jerusalén. Un estudio sobre la banalidad del mal*. Barcelona: Lumen.

Balzarini, M. (2023a). ¿Prevención del acto suicida? In *Revista Psicoanálisis en la universidad* N°7. Rosario, Argentina, UNR Editora, pp. 145–153. Accessed 2 June 2023 from: https://psicoanalisisenlauniversidad.unr.edu.ar/index.php/RPU/article/view/ 155/118

Bassols, M. (2011a). *Tu yo no es tuyo*. Buenos Aires: Tres Haches.

Bassols, M. (2011b). Las neurociencias y el sujeto del inconsciente. Conferencia pronunciada en Granada. Instituto del Campo Freudiano. Accessed from: www.icf-granada. net/2012-04-04-08-33-03/videos/83-las-neurociencias-y-el-sujeto-del-inconsciente

Bassols, M. (2012a). Lo real del psicoanálisis. Lo real en la ciencia y el psicoanálisis. *Virtualia* (25). Accessed 17 July 2020 from: www.revistavirtualia.com/articulos/262/ lo-real-en-la-ciencia-y-el-psicoanalisis/lo-real-del-psicoanalisis

Bassols, M. (2012b). Psicoanálisis, sujeto y neurociencias. Presentación del libro *Sutilezas analíticas* en Alianza Francesa de San Ángel. Nueva Escuela Lacaniana. Mexico D.F.

Bassols, M. (2012c). Lo real del psicoanálisis. In *Freudiana* (64). ELP de la EFP miembro de la AMP. Catalunya: Repro Disseny.

Bassols, M. (2013a). La vigencia del psicoanálisis. Entrevista por Bordon, J.M. en Revista *Noticias* (pp. 118–120). Centro de investigación y docencia en psicoanálisis. Lima. Accessed from: www.enapol.com/images/Prensa/13-12-06_Entrevista-a-Miquel-Bassols. pdf

Bassols, M. (2016a). Freud era un misógino contrariado, pero se dejó enseñar por las mujeres. Entrevista en *El País* por Ángela Molina. Accessed from www.eol.org.ar/template.asp?S ec=prensa&SubSec=europa&File=europa/2016/16-03-27_Entrevista-a-Miquel-Bassols. html

Bassols, M. (2016b). La fascinación mecánica. In *Freudiana* (77–78). ELP de la EFP miembro de la AMP. Catalunya: Repro Disseny.

Bassols, M. (2016c). La "substancia gozante". In *Revista Lacaniana* (21). El racismo que me habita. Buenos Aires: Escuela de la Orientación Lacaniana.

Bassols, M. (2019). El falso dualismo entre mente y cuerpo. In *Freudiana* (86) "Inconsciente y cerebro: nada en común". ELP de la EFP miembro de la AMP. Catalunya: Repro Disseny.

Bassols, M., Laurent, E., and Berenguer, E. (2006). Lost in cognition. In *Freudiana* (46). ELP de la EFP miembro de la AMP. Catalunya: Repro Disseny.

Belaga, G. (2011). La salud mental, lo inevitable de una totalidad fallida. In *Lacaniana de psicoanálisis* (11), año VII, pp. 41–44. Publicación de Escuela de Orientación Lacaniana. Buenos Aires: Grama.

Briole, G., Rouse, H., Fernández, M., Galaman, C., Godínez, R., and Teixidó, A. (2019). Argumento tercera reunión "La clínica bajo transferencia: nada que ver con los tratamientos neuro". Prólogo al 5° Congreso de la EuroFederación de Psicoanálisis "PIPOL9". Inconsciente y cerebro: nada en común.

Brodsky, G. (2013). Contra la ilusión religiosa. La unión solidaria del psicoanálisis y la ciencia según Freud. Publicación de Escuela de Orientación Lacaniana. Accessed from: www. eol.org.ar/template.asp?Sec=prensa&SubSec=america&File=america/2013/13-08-22_ Contra-la-ilusion-religiosa.html

Brodsky, G. (2015a). Mi cuerpo y yo. Conferencia pública. Participan Universidad del Claustro de Sor Juana, local de la NEL-México DF y Alianza Francesa de San Ángel. México. Accessed 9 November 2019 from: www.radiolacan.com/es/topic/589/8#. XUQ10i5E2oU.whatsapp

Camerer, C., Loewenstein, G. y Prelec, D. (2005). Neuroeconomics: how neuroscience can inform economics. *Journal of Economic Literature*, XLIII, 2005, pp. 9–64.

Cancina, P. (2008). *La investigación en psicoanálisis*. Argentina: Homo sapiens.

Castanet, H. (2023). *Neurología versus psicoanálisis*. Buenos Aires: Grama Navarin.

Chamorro, J. (2011). *¡Interpretar!* Buenos Aires: Grama.

Cosenza, D. and Puig, S. (2019). La tentación neurobiológica del psicoanálisis y el corte de Lacan. Presentación Hacia Pipol 9: El inconsciente y el cerebro: nada en común. Escuela Lacaniana de Psicoanálisis de Catalunya del Campo Freudiano. Accessed 15 November 2020 from: www.cdcelp.org/es/ficha-actividad.php?f=396&s=1

Dall'Aglio, J. (2020a). No-Thing in common between the unconscious and the brain: on the (im)possibility of Lacanian Neuropsychoanalysys. En ResearchGate, Psychoanalysis Lacan, 4. Accessed 15 April 2023 from: https://researchgate.net/publication/342870600

Dall'Aglio, J. (2021). What can psychoanalysis learn from neuroscience? A theoretical basis for the emergence of a neuropsychoanalytic model. En Contemporary psychoanalysis, 57 (1), 125–145. doi: 10.1080/00107530.2021.1894542

De Georges, P. (2005). Paradigma de desencadenamiento. In Jacques-Alain Miller et al., *Los inclasificables de la clínica psicoanalítica*. Buenos Aires: Paidós, pp. 41–46.

Dessal, G. (2016). Vicisitudes actuales de la vida amorosa. Conferencia. Accessed 25 May 2023 from: www.youtube.com/watch?v=nGUxYJVc58U

Esqué, X. (2017). ¿Por qué analizarse hoy? Conferencia Facultad de Psicología, UNC. Córdoba.

Freud, S. [1916] (1986b). Una dificultad del psicoanálisis. In *Sigmund Freud. Obras Completas*. Tomo XVII. Buenos Aires: Amorrortu.

Freud, S. [1940] (2006g). Esquema del psicoanálisis. In *Sigmund Freud. Obras Completas*. Tomo XXIII. Buenos Aires: Amorrortu.

Freud, S. [1915] (2012b). Lo inconsciente. In *Sigmund Freud. Obras Completas*. Tomo XIV Buenos Aires: Amorrortu.

García de Frutos, H. (2011). El psicoanálisis no es una ciencia. In *Freudiana* (62). ELP de la EFP miembro de la AMP. Catalunya: Repro Disseny.

García de Frutos, H. (2012). Neurocientificismo, logicismo y psicoanálisis: algunos apuntes para una perspectiva crítica. In *Freudiana* (65) "Los espectros del autismo". ELP de la EFP miembro de la AMP. Catalunya: Repro Disseny.

Han, B.-C. (2012). *La sociedad del cansancio*. Barcelona: Herder.

Han, B.-C. (2014). *La agonía del Eros*. Barcelona: Herder.

Han, B.-C. (2022). *Capitalismo y pulsión de muerte*. Barcelona: Herder.

Handke, P. (2006). *Ensayo sobre el cansancio*. Madrid: Alianza.

Jones, E. (1981). *Vida y obra de Sigmund Freud*. Tomo 1. Barcelona: Anagrama.

Kandel, E. (2009a). Aspiraciones de la biología para un nuevo humanismo. In E. Kandel (ed.), *Psiquiatría, psicoanálisis, y la nueva biología de la mente*. Tercera ed. España, Barcelona: Ars Medica.

Koretzky, C. (2019). *Sueños y despertares. Una elucidación lacaniana*. Buenos Aires: Grama

Lacan, J. [1974] (2006b). *El triunfo de la religión*. Buenos Aires: Paidós.

Lacan, J. [1969–1970] (2008e). *El Seminario. Libro 17. El Reverso del Psicoanálisis*. Buenos Aires: Paidós.

Lacan, J. [1972–1973] (2008f). *El Seminario. Libro 20. Aun*. Buenos Aires: Paidós.

Lacan, J. [1965] (2009a). La ciencia y la verdad. In *Escritos 2*. Buenos Aires: Sigloveintiuno.

Lacan, J. [1955] (2009c). La cosa freudiana o sentido del retorno a Freud en psicoanálisis. In *Escritos 1*. Buenos Aires, Argentina: Sigloveintiuno.

Lacan, J. [1953] (2009d). Función y campo de la palabra y del lenguaje en psicoanálisis. In *Escritos 1*. Buenos Aires, Argentina: Sigloveintiuno.

Lacan, J. [1951] (2009e). Intervención sobre la transferencia. In *Escritos 1*. Bs Aires: Sigloveintiuno.

Lacan, J. [1971–1972] (2012a). *Hablo a las Paredes*. Buenos Aires: Paidós.

Lacan, J. [1971–1972] (2012c). *El Seminario. Libro 19. [...] o peor* Buenos Aires: Paidós.

Lacan, J. [1964] (2013). *El Seminario. Libro 11. Los cuatro conceptos fundamentales del psicoanálisis*. Buenos Aires: Paidos.

Langaney, A. (2006). El sentido de la seducción. In *Mente y cerebro. Freud. Investigación y ciencia* (18), 80–82.

Lardjane, R. (2019). El inconsciente y el cerebro en psiquiatría. In *Lacan cotidiano. Para Pipol 9*. Revista de Psicoanálisis (824). BOLC.

Laurent, E. (2002). *Síntoma y nominación*. Buenos Aires: Diva.

Laurent, E. (2005). *Lost in cognition. El lugar de la pérdida en la cognición*. Buenos Aires: Diva.

Laurent, E. (2014b). El psicoanálisis no es una psicoterapia, pero […] In *Freudiana* (70). ELP de la EFP miembro de la AMP. Catalunya: Repro Disseny.

Laurent, E. (2016a). *El reverso de la biopolítica*. Buenos Aires: Grama.

Laurent, E. (2020a). Hay tantas fes. *Revista Lapso* (5). Maestría en teoría psicoanalítica lacaniana. UNC. Accessed 28 August 2020 from: http://matpsil.com/revista-lapso/portfolio-items/laurent-video-videoentrevista/

Laurent, E. (2020b). El nombre y la causa. Conicet y UNC. Córdoba: IIPsi Instituto de Investigaciones Psicológicas.

Manes, F. (2019). Seis consejos para cuidar la salud de tu cerebro. Accessed from: https://aprendemosjuntos.elpais.com/especial/la-vida-no-es-la-que-vivimos-sino-como-la-recordamos-para-contarla-facundo-manes/ Accessed from www.youtube.com/watch?v=3-18pPudCxM&feature=youtu.be

Martinez, M. (2018). Cuando la ciencia vacila. Entrevista realizada por Silvia Salvarezza y Luis Martínez. In *Revista Lacaniana de Psicoanálisis* (24), Ciencia Ficción, EOL, 13. Buenos Aires, Grama, 193–195.

McCabe, D. and Castel, A. (2008). Seeing is believing: the effect of brain images on judgments of scientific reasoning. *Cognition*, 107, 343–352.

Miller, J. (2018). Cientismo, ruina de la ciencia. In *Revista Lacaniana de Psicoanálisis* (24), Ciencia Ficción, EOL, 13. Buenos Aires, Grama, pp. 11–13.

Miller, J.-A. (1987a). "Entrevista con Jacques-Alain Miller" por Gonzalez, F. In *Revista A.E.N.* Vol. VII. Nº 23.

Miller, J.-A. (1994b). Psicoterapia y psicoanálisis. In *Revista Freudiana* (10). Escuela Europea de Psicoanálisis-Catalunya.

Miller, J.-A. [1986–1987] (1998a). *Los signos del goce*. Buenos Aires: Paidós.

Miller, J.-A. [1981] (1998d). Psicoanálisis y psiquiatría. In *Elucidación de Lacan. Charlas brasileñas*. Buenos Aires: Paidós.

Miller, J.-A. (2004c). Verdad, probabilidad estadística, lo real. In *Revista Lacaniana. Las prácticas de la escucha y sus argumentos* (2). Buenos Aires: EOL.

Miller, J.-A. (2005). Cierre. Vacío y certeza. En *Los inclasificables de la clínica psicoanalítica*, Buenos Aires, Paidós, pp. 189–193.

Miller, J.-A. [1985–1986] (2010). *Extimidad*. Buenos Aires: Paidós.

Miller, J.-A. [1999] (2011c). Biología lacaniana. In *La experiencia de lo real en la cura psicoanalítica*. Buenos Aires: Paidós.

Miller, J.-A. [1988] (2012a). El psicoanálisis, su lugar entre las ciencias. In *Revista consecuencias* (9). Accessed 17 May 2020 from: www.revconsecuencias.com.ar/ediciones/009/template.php?file=arts/Alcances/El-psicoanalisis-su-lugar-entre-las-ciencias.html

Miller, J.-A. [2000–2001] (2013b). *El lugar y el lazo*. Buenos Aires: Paidos.

Miller, J.-A. [2008–2009] (2014a). *Sutilezas analíticas*. Buenos Aires: Paidós.

Miller, J.-A. [2008] (2015a). *Todo el mundo es loco*. Buenos Aires: Paidós.

Miller, J.-A. [1980] (2015b). *Seminarios de Caracas y Bogotá*. Buenos Aires: Paidos.

Miller, J.-A. (2016b). ¿Ha dicho raro? In *Revista Mediodicho* (42) ¿A qué le tenemos miedo? Publicación de EOL. Córdoba, Argentina.

Miller J.-A., (2018c). Neuro-, le nouveau réel. In *La Cause du désir* (98).

Miller, J.-A. [1984–1985] (2021d). *1, 2, 3, 4*. Buenos Aires: Paidós.

Milner, J.-C. (1996). *La obra clara. Lacan, la ciencia y la filosofía*. Buenos Aires: Manantiales.

Moraga, P. (2019). El paradigma forclusivo de las neurociencias. In *Lacan cotidiano. Para Pipol 9*. Revista de Psicoanálisis (824). BOLC.

Ordóñez, P. (2023). Año nuevo: ¿una nueva vida para un nuevo tiempo? In *Comercio y justicia*, 84 (24041), periódico de Ciudad Autónoma de Buenos Aires, 2 January 2023.

Peteiro, J. (2010). *El autoritarismo científico*. Colección Itaca.Málaga, España: Miguel Gomez.

Pinker, S. (2004). Órganos de computación. Entrevista realizada por Brockman. In *Revista Lacaniana. Las prácticas de la escucha y sus argumentos* (2). Buenos Aires: EOL.

Pino, S. (2018). La utilidad de las ficciones. In *Revista Lacaniana de Psicoanálisis* (24), Ciencia Ficción, EOL, 13. Buenos Aires, Grama, pp. 50–53.

Ritvo, J. (2014). *La retórica conjetural o el nacimiento del sujeto*. Rosario: Nube Negra.

Rodriguez, T. (2013). Furtivos saboteadores de la salud. In *Mente y cerebro. El legado de Freud*. La neurociencia demuestra la eficacia del psicoanálisis. *Investigación y ciencia* (62), 5.

Rose, N. (2011). Gobernar la conducta en la época del cerebro. Conferencia dictada en III Coloquio Latinoamericano de Biopolítica, Universidad Pedagógica (UNIPE). Buenos Aires.

Slezak, D. (2018). Una App que ayuda a diagnosticar esquizofrenia a través del análisis del discurso de pacientes. Conicet. UBA. Argentina. Accessed from: www.conicet.gov.ar/una-app-que-ayuda-a-diagnosticar-esquizofrenia-a-traves-del-analisis-del-discurso-de-pacientes/

Solms, M. (2006). Neuropsicoanálisis. Interview by Steve Ayan. *Mente y cerebro. Freud. Investigación y ciencia* (18), 74.

Stagnaro, J. (2009). "Psiquiatría y neurobiología: el arte de curar y la ciencia del cerebro en crisis paradigmática". In *Jacques Lacan y los matemáticos, los lógicos y los científicos*. Buenos Aires: Escuela Freudiana de Buenos Aires.

Stecco, C. (2020). El ombligo del sueño. Un impoético. In *Scilicet* El sueño. Su interpretación y su uso en la cura lacaniana. Publicación en razón del XII congreso de la Asociación Mundial de Psicoanálisis. Buenos Aires: Grama.

Teixidó, A. (2019a). Des-cerebrados. In Prólogo al 5° Congreso de la EuroFederación de Psicoanálisis "PIPOL9". Inconsciente y cerebro: nada en común. Accessed 30 May 2020 from: www.cdcelp.org/es/ficha-actividad.php?f=396&s=1

Ubieto, J. (2019b). El paradigma "neuro" y las paradojas del goce. In *Prólogo* hacia el 5° Congreso de la EuroFederación de Psicoanálisis "PIPOL9". Inconsciente y cerebro: nada en común. Accessed 30 May 2020 from: www.cdcelp.org/es/ficha-actividad.php?f=396&s=1

Ubieto, J. (2019c). El paradigma "neuro" y las paradojas del goce. In *Freudiana* (86) "Inconsciente y cerebro: nada en común". ELP de la EFP miembro de la AMP. Catalunya: Repro Disseny.

Vilá, F. (2019a). La ideología de la Ciencia, la muerte y el sexo. En Prólogo hacia el 5° Congreso de la EuroFederación de Psicoanálisis "PIPOL9". Inconsciente y cerebro: nada en común. Accessed 30 May 2020 from: www.cdcelp.org/es/ficha-actividad.php?f=396&s=1

Yellati, N. (2018). *Lo que el psicoanálisis enseña a las neurociencias*. Buenos Aires: Grama.

Zack, O. (2007). *Un uso ético de los antidepresivos. Entrevista para El Tribuno*. Salta: EOL.

Chapter 5

Neuroscientific imputations to psychoanalysis

Therapeutic ineffectiveness

We know that psychoanalysis is the subject of various accusations, including that it is ineffective. Psychoanalysis, the practice of listening, is being branded obsolete. Freud is rejected saying that he built a theory from the single case and therefore has no scientific validity.

Such rejection feeds the tendency toward therapeutic demand, toward direct utility, toward immediate efficacy. If you go to the dentist, you expect him to cure you; one thinks, is he going to cure me? If there is something that has changed, if something was there and no longer bothers, there is something therapeutic there. The therapeutic thing is to remove what bothers. There's a utility dimension to this if something has changed; it was bugging you and it's gone. We must say that the neurotic formula is "I want to, but I can't". If we go along the side of helping so that he can, because we know the organic locality on which to operate to help make the desire possible, if we go along that path of helping him with what he can't do, but wants, we get into the therapeutic demand. That is why psychoanalysts question the "I want". If he says: "I have a problem, I feel alone", what is the problem of feeling alone? They question what is true so that the analysis session is a space for breathing.

Mías (2008), an Argentine neuropsychologist, points out that psychoanalysis requires prolonged treatments and this

> from the methodological point of view does not allow us to specify whether the results are the product of the treatment, of the natural evolution-maturation of the subject in the face of external events, or of an interactive effect between the two.
>
> (pp. 37–38)

He says that "psychoanalysis has generated hypotheses, rather than compared them. As a result, it has not been able to evolve as other fields of psychology and medicine have" (p. 70). For their part, Buchheim et al. (2013) affirm that the analytical effects are demonstrated with brain tests: "After a few months of psychoanalysis,

DOI: 10.4324/9781003458470-6

various brain areas of the patients reacted less markedly to certain phrases that, before treatment, activated adverse attachment patterns" (p. 29). Likewise, Solms (2006) claims that Freud's weak point was that the method he created had nothing to do with scientific verification: "Psychoanalysis certainly elaborates more conjectures about psychic processes than can be deduced only from observation of behavior [...] empirical research can do very little in this field" (p. 74). Thus, neurosciences seek to "activate unconscious processes in order to record them using neuroscientific techniques" (Buchheim et al., 2013, p. 29), and with this they argue that "[...] the intersection of neurobiological research and psychoanalysis is fertile and growing. That would have delighted Sigmund Freud" (Delgado, Strawn, and Pedapati, 2015, p. 39).

To be strong in this world you have to demonstrate effectiveness. For these neurosciences, the weak point of psychoanalysis is its lack of efficacy. Therapeutic efficacy is associated with the test of truth. Laurent (2014b) points out that the psychoanalyst is not trained to be a psychotherapist, this occurs "in addition", that is, the therapeutic effects in a psychoanalysis occur as a consequence. However, today everything has to be instant, it has to be effective, and everything has to be useful, while the psychoanalyst does not offer himself as a useful tool, that is what electrical appliances are for. Psychoanalysis does not go down that path, of Jeremy Bentham's utilitarianism (Lacan, 2008f), but is a practice that is offered not to be effective, but so that the subject can live a little better.

If it is a question of explaining in which part of the brain is the suffering that the subject does not know he is carrying, psychoanalysis is useless for that. But if you have to answer what is the psychoanalysis for, he is used so that the subject can choose a partner to whom he can tell his own shit (Grinbaum, 2020). For example, a patient says: "the truth, today you told me nothing. This is useless to me", the analyst: "Let's look for the useless character in your life, which is me in this case, I'm useless!" (Chamorro, 2011, p. 62). The psychoanalyst serves in the position of an object because the one who demands is the subject, not the analyst. As Miller (2016a) says, the subject demands to know something about his jouissance. If the analyst is in the position of the analyst's desire, he does not demand, he does not mean, he abstains. He promotes that the subject's demands are placed as a cause of the subject meaning something. It is a lodging, which is not lodging the unconscious in the brain, but in love.

The analyst does not call the patient because he has not come for a long time, he does not require him to continue the analysis, and he does not force him beyond his happiness, which is the law in some "chronic" treatments. The analyst accompanies, even when the subject distances himself from the analysis, when he begins to get exhausted, he forgets to go, he says that he is tired, that he was glad to meet, for example, with his family, that he stayed drinking late and forgot the session, that he took a nap and the hour passed, that he had a dance class, in short, he begins to libidinize his world and forgets about the analysis, that is what should happen (Chamorro, 2011). Instead, neurosciences are offered as what cannot be discarded, the highest good with which pleasure is obtained because the therapist knows how to adjust the subject to reality. Psychoanalysts do not know *a priori* what is best for a subject, who should sleep with him, and what should he study. Thus, what makes a psychoanalyst Lacanian is a privilege: that of having gotten rid of all prior judgment (Grinbaum, 2020). That is to say, the only effectiveness of psychoanalysis is to be a discourse that does not seek to dominate.

No falsifiability

Psychoanalysis is accused of not being scientific because it is not refutable, that is to say that it is always right. According to Popper's (2001) fallibility criteria, for a practice to be considered a science it must be able to be falsifiable, it must be capable of formulating its own refutation procedure, "of defining the conditions such that, if they are not respected, the discourse concedes that is false" (p. 46). Popper affirms that psychoanalysis cannot be a science because if it cannot define under what conditions of experience the hypothesis would be recognized as false, it cannot be refutable, then it cannot be scientifically proven. It is something that Lacan grants to Popper. Freud's first theory could not be demonstrated, it was "the case proves the theory". The point is that Freud refutes his first theory, which results in the second topic. Starting from *Beyond the Pleasure Principle*, he recasts the theory of drives to include the death drive, from which it follows that psychoanalysis was contrasted by Freud himself (Regnault, 2004; Miller, 2021b, 2021d). Lacan also went against himself; he began his teaching "the primacy of the symbolic" while in the 1970s we have "RSI", Borromean knot, there is no primacy of one register over another, if one is unhooked, they are all untied, and the three registers are now on the same level. So, is it true that psychoanalysis has not falsified its hypotheses?

We also have the effects of the great cases of Freud and Lacan, the testimonies of the past where the AEs ["Analista de Escuela", in English "School Analyst", is the highest nomination granted by the member Schools of the World Association of Psychoanalysis to analysts who have presented themselves to what Lacan invented as a guarantee instrument and called the device of the pass and have managed to demonstrate their end of analysis before several instances of evaluation] have given an account of how psychoanalysis allowed that subject to find a way to manage with his jouissance, the AME nomination ["Analista Miembro de la Escuela", in English "Analyst Member of the School" is one of the nominations granted by the member Schools of the World Association of Psychoanalysis to practitioners who have requested their membership and, after having passed instances of evaluation, have demonstrated a history of sustained work in relation to the cause of psychoanalysis and in relation to the School, which have realized that they are in analysis and that they control their practice] that supposes a desire decided to work for the cause and survival of psychoanalysis – to work not for the analyst, but for the field that concerns psychoanalysis – and the effects that we feel in ourselves, then, as analysands. After all this, does it need to be empirically provable?

All in all, as much as psychoanalysis resists criticism regarding its non-scientific nature, the speaking being requires a listening that is divorced from the verificationist method. If Lacan conceded to Popper the non-scientificity of psychoanalysis, it is because the "hypothesis of the unconscious fits very well to saying that it does not exist" (Miller, 2021d, p. 48). This does not faze psychoanalysts. It is even what Lacan formulates, the unconscious on the side of what is not, of non-being, since its being is all in potentiality (Miller, 2021d).

That psychoanalysis cannot be contrasted, that falsification cannot be applied to it, that is why it is not scientific, is an argument that has been extended to the present. The attempt to codify the good practices of psychologists, of psychiatrists, in the decrees that legislate the actions of health professionals, in hospital services, in rehabilitation centers, recall the incessant attacks against psychoanalysis. Privileging neurocausality leads to the understanding that suffering is an anomaly in the normal-ideal course of human evolution. The psychologist must quantify this suffering with neuropsychological tests and thus may receive fees for his good practice. "They want to place the clinical practices of the word under the authority of organic medicine and demand that they be oriented exclusively from the neurobiological thesis" (Castanet, 2023, p. 26). Good clinical practice protocols are obtained that depend on this hilarious bureaucratic-scientific style, which designates disorders to mobilize combos of psychopedagogy sessions, phonoaudiologists, psychomotor skills, kinesiology, support teachers, occupational therapists, and other specialized professionals. For some reason the State, through laws, invests its budget in this. It is at the forefront of neuro-detection and methods (Castanet, 2023).

It's not a science

The assertions that psychoanalysis is a science or psychoanalysis is not a science do not account for the complexity involved. Whether Lacan avoided those demarcations is not the way the matter should be posed. In fact, what Lacan contributes is not reduced to the passage of psychoanalysis from a natural science to a conjectural science because, being a science, it does not eliminate the notion of cause that is typical of the search for meaning. Nor is it taking psychoanalysis to linguistics because this is a science of etiology, of the archaeology of language that places the subject in the causal relationship as the subject of the signifier. Lacan's essential contribution is to retake Freud's saying, subsumed to the axiom there is no sexual relationship, which is, as Miller (2019a) indicates, what gives reason to the analytic experience and which could be qualified as a disproportion between cause and effect. Between cause and effect there is emptiness, not causality.

> In the physical order, the progress of science shows us that we are in a position to control the cause-effect relationship, but what remains opaque and problematic is knowing how a physical causality touches the psychic and how the psychic could be inserted into the causality physical.
>
> (Miller, 2019b, p. 75)

That a part of the body reacts does not mean that there is a response in subjective terms. The reaction follows a tension-elevating stimulus, while the response involves a more complex structure. A reaction is not a response because the meaning of a subjective experience cannot be captured with the neuronal biological. The reaction is necessary to have answers, but the sufficient condition for them is language (Bassols, 2011b).

> The complex synapses that the neurologist discovers in the brain, the astonishing network that constitutes this object, the most evolved in the universe, cannot account for this interval [...] Nothing of me can be found in the brain [...] because "me" is a word whose code is not reducible to that of the cerebral organization.
>
> (Ritvo, 2014, p. 27)

There are no laws of the relationship between the subjective and the brain. "For this very reason, the discourse of analysis is distinguished from scientific discourse" (Lacan, 2008f, p. 141).

It is true that for Lacan psychoanalysis shares with science a horizon characterized by the need for formalization, but as regards its inclusion or not in the field of science, the approach is different. In 1964, with the status of the pulsating unconscious and its inherent evanescence, he pointed out a direction that distanced the unconscious from its scientific nature. In the inaugural class of that year's seminar, the question "what is psychoanalysis?" leads Lacan to wonder if psychoanalysis is a science or, in other words, what would be required to authorize psychoanalysis to call itself a science. As long as psychoanalysis is about a praxis (human action to treat the real through the symbolic) it must be distinguished from the question of knowing whether or not it is a science (Lacan, 2013). A praxis depends on what happens in the experience; he does not seek the theory to be verified, but rather the surprise to take place to produce an encounter. It is not an applied technique, but rather a discourse that encourages everyone to produce their singularity, their exception (Laurent, 2004).

Then, in 1965, Lacan is no longer concerned with whether psychoanalysis is a science, but with how science should be reformulated if it claims to include psychoanalysis: "Then, the question that constitutes our radical project remained permanent, the one that goes from: is psychoanalysis a science? a: what is a science that includes psychoanalysis?" (2012h, p. 205). Years later, in 1973 he asked: "How is a science still possible after what can be said of the unconscious?" (2008f, p. 127), a question that expresses his disbelief about the possibility that science can be restructured if it tries to work with the unconscious. Finally, in 1974 he was no longer concerned about how to include psychoanalysis in science. "The problem that we are faced with is no longer that of knowledge or of 'co-birth', that of a 'co-naturalness' by which we are entrusted with the friendship of appearances" (2006b, p. 49). Indeed, psychoanalysis does not claim to be a science, despite being born with it (García de Frutos, 2011).

Diffusion deprivation

Kandel (2009a, 2009b) accuses psychoanalysis of not accounting for the interventions it carries out, of making them exclusive, and of depriving the dissemination of the data it collects. And the pass device? And the congresses, conferences, etc.? Of course, such meetings respect the privacy of the clinic, since it is from there that an interpretation that is never reproducible is built. However, Kandel says:

> With few infrequent exceptions, the data collected in psychoanalysis sessions are private: the patient's comments, associations, silences, postures, movements and other behaviors are confidential [...] And this is precisely the problem.
>
> (Kandel, 2009b, p. 70)

Kandel affirms for this reason that psychoanalysis has fallen into decline.

> The most important thing, and the most disappointing thing, is that psychoanalysis did not evolve scientifically. Specifically, it did not develop objective methods for testing the interesting ideas he had formulated. As a result, psychoanalysis entered the 21st century with its influence on the wane.
>
> (p. 389)

It is true that Freud had contracted obligations not only toward the care of the sick, but also toward science. But the public communication of what he thought he knew about the cause of the hysteria was something he expressly kept in reserve. The privacy of the psychosexual life of the patients and the expression of their secret repressed desires was guarded with absolute prudence by Freud (2011ll):

> I know that there are – at least in this city – many doctors who (quite disgustingly) will want to read a clinical case of this nature as a novel with a code intended for their amusement and not as a contribution to the psychopathology of neuroses. To this class of readers, I assure you that any case histories I may publish in the future will circumvent their sagacity by similar guarantees of secrecy, even if this purpose forces me to restrict myself enormously in the use of my material.
>
> (p. 8)

Freud restricts the use of the material because what occurs in one session is unique to that person's experience and cannot be applied to others. Although he publishes Dora's clinical material, he does so in a specialized magazine to protect the patient from the harm that could have been caused by non-specialists learning about her secrets. Thus, Freud and science point to the publication of the clinical case with the difference that Freud was moved by the cause of desire while science was moved by the validation of the administration of the technique.

There is something in this society that also goes against privacy. The thing is displayed directly, no detours are taken, distances are shortened, allusions have disappeared, and it is the spectacle of intimacy. "Today we can no longer stand the slow, the long, the silent. We no longer have patience" (Han, 2022, p. 116). Concealment, distraction, "withholding information, is no longer tolerated" (pp. 131–132). This is what Kandel complains about: psychoanalysis retains the private. And it's current! Today everything has to be visible. As Han (2022) states, it is a society of transparency, whose threat is that it has become a society of control. The countless surveillance cameras are suspicious of each one of us. The body is the instrument of the digital panopticon that promulgates the imperative "you have to be free", translated as "you must hide nothing" which means "nothing to decipher". It is a domination that encourages us not to protect ourselves. Today we do not object to their being collected, data about our consumption behavior, marital status, and people with whom we maintain relationships are stored, and known, everything is recorded, each step we take is reconstructable, and each click we make is recorded, our habit is reproduced exactly on the web, an external brain that thinks for us. Data is public, transparent, and becomes controllable. Everything is known, ignorance has been eliminated, interpretation has died, and trust is not needed if information can be easily obtained. Clinic of the unbearable of the secret, where the absolutization of visibility and the certainty of transparency make the object more essential than its veil. This leaves aside the Freudian demonstration that the human being is not transparent even to himself. Freud reads Goethe's Faust where a beautiful thing is appreciated for what it does not show. Beauty is concealment, the impossibility of the thing being revealed. That idea of beauty, which eroticizes, which manages not to show the naked truth and invites curiosity, is totally abolished by these neurosciences that promulgate the resignation of private refuges.

The borders inside and outside are today permeable thanks to plasticity, to interconnections. It is in line with the big data fever, the rage to collect data, control subjects and drive them to performance. It should not be ignored that neurosciences have been seduced by this dataism. For example, in some relaxation practices sensors are applied that automatically measure all body parameters, temperature, steps, sleep cycles, calorie intake, movement profiles, and brain waves. Even in meditation, the pulsations are being recorded according to the protocol. "So, even in relaxation, performance and efficiency count, which is actually a paradox" (Han, 2022, p. 65).

Universe of discourse

Psychoanalysis is accused of not having incorporated knowledge from other disciplines and of having fallen into the air of a universe of discourse. Let's read Freud:

> We have often heard the claim that a science must be built on clear and precisely defined basic concepts. In reality, none, not even the most exact, begins with such definitions. The correct beginning of scientific activity consists rather in describing phenomena that are then grouped, ordered and inserted into connections.
>
> (2012g, p. 113)

Freud was clear that a science cannot define alone and by itself the concepts that support its practical operation. For his part, Lacan considered that concepts did not have to be jealously guarded by the disciplines that created them. Among the disciplines related to psychoanalysis, which according to Lacan contributed to psychoanalytic theory, are linguistics, logic, topology, and philosophy under the modality of anti-philosophy. But not biology.

Insufficiency in the explanation of the mind

Psychoanalysis is accused of explaining the mind in a vague way. Does psychoanalysis explain? It is true that the analyst is asked to say something, to direct the associations, otherwise what he directs is the madness of the patients who continue with their symptoms if the analyst remains silent. It is true that the analyst is asked to conclude, but this is too different from speaking. Didn't Freud say that the patient is cured by talking? Bertha Pappenheim (cit. Jones, 1981) reproached Breuer: "let me speak doctor". The speaker is in a position to explain. The logic of exchange typical of capitalism indicates "if I pay you, you have to give me an explanation". Does this complicate psychoanalysis? According to Kandel (2009a), the patient has to learn, therefore the therapist would have to explain.

Postpone relief

Psychoanalysis is accused of postponing the relief of the person (Kandel, 2009a). In this age where advances in technique are guided by the premise "there is no pain", which is another way of saying "be happy", it should be remembered that Freud located an intrinsic malaise in culture. The pain does not go away. Perhaps the psychoanalyst is judged as a masochist, old, unscientific, but it leads to a real life.

It does not submit its operation to expert evaluation

Kandel objects that psychoanalysis is not concerned with what others may judge about the interventions that occur in each session, claiming that it would have to go through a committee to grant credibility, validity, and methodological reliability and thus prevent the intervention from being biased by the solitary judgment of the analyst: "[…] psychoanalysts are rarely concerned that their view of what has occurred in a therapy session is biased or subjective" (2009b, p. 70). Is it with a subjective bias that a psychoanalyst works or is it with the training that he can refrain from answering in the name of his ghost? Did Freud teach that the reading of the subjective text must be subjected to a judgment that evaluates the deviation according to factual reality, which has now become physical reality, or did he teach that in order to prevent the expert's judgment from getting in the way of the cure, which today became a neuroscientist judgment, and psychic reality is prioritized, must the analyst have undergone a psychoanalytic purification? We wonder, have these eminent neuroscientists submitted to psychoanalysis? It would be important that they give proof of this. Asking for evidence is not unknown to these neurosciences.

Putting something to the test constitutes the discourse they hold. What is, for neuroscience, a hypothesis that cannot be tested?

The non-existence of the sage

The science of calculations does not accept that knowledge remains unknown, so the science of calculations does not accept the unconscious (Castanet, 2023). The challenge of scientific discourse is to ensure the conversion of unknown knowledge to transmittable and quantifiable knowledge, but who can make anxiety enter the variables of the quantification regime? For that is the structured interview: "a lot, sometimes, a little, not at all, normal delivery, caesarean section, hypoxia x months of pregnancy, lactation period", mark with a cross! Complete anamnesis, where the questions about the patient's history become questions about the family history; hours of work to clear that interview full of information; more than two professionals working, more than one unconscious, because the whole family comes to the consultation; and if the patient continues with the symptoms they are chronic patient.

Where is all this going? Toward not wanting to know about the non-existence of the sage. Laurent (2011) warns of the anguish that the neurobiologist goes through. By the nature of truth the wise does not exist. And not because the neuroscientist has not made training efforts. The tendency to make the wise exist, indicates Laurent (2005), is sustained by the push that guilt produces under the unbearable idea of having failed. That happens when it is intended to objectively demonstrate the effectiveness of the act. On the contrary, Castanet (2023) indicates that psychoanalysis "does not have to clear its name or render accounts. Although it is not a hard science, it nevertheless chooses rigor and logical demonstration" (p. 34).

It is not welcome!

The knowledge of psychoanalysis, which is obtained from a personal psychoanalysis, is little taken into account in educational, legal, governmental spaces and not because it does not contribute, in fact its contribution is in questions, but because, instead of speeding up responses and saving work for the consultant, it stops so as not to privilege behavior and adaptation.

As Castanet (2023) says, the current attacks against psychoanalysis reveal a logic that Miller cleared up in 2004: "The course of civilization leads to the suppression of psychoanalysis" (p. 24). The attempt to assassinate psychoanalysis, as shown by Aflalo (2011), to put an end to the clinic of the word, clearly reveals that psychoanalysis deals with anguish. And it is there that it becomes the object of reproach for a world that changes rapidly. What would have happened if the assassination attempt on psychoanalysis had been successful? Perhaps we would return to the obscurantism, to magic, and to the hysterical pathologies of two centuries ago.

These accusations are evidence that psychoanalysis always developed under constant threat of extinction. From the start the standards do not mesh with the cutting edge of Freudian truth. Psychoanalysis always goes against the current and for this reason it is not welcome (Esqué, 2017). The power structures have never rolled out the red carpet. The psychoanalyst has never lamented for this. And if it ever happens, he should ask himself, what has he done to deserve it?

References

Aflalo, A. (2011). *El intento de asesinato del psicoanálisis*. Buenos Aires: Grama.

Bassols, M. (2011b). Las neurociencias y el sujeto del inconsciente. Conferencia pronunciada en Granada. Instituto del Campo Freudiano. Accessed from: www.icf-granada.net/2012-04-04-08-33-03/videos/83-las-neurociencias-y-el-sujeto-del-inconsciente

Buchheim, A., Cierpka, M., Kächele, H., and Roth, G. (2013). Efectos del psicoanálisis en el cerebro. In *Mente y cerebro. El legado de Freud*. La neurociencia demuestra la eficacia del psicoanálisis. *Investigación y ciencia* (62), 26–29.

Castanet, H. (2023). *Neurología versus psicoanálisis*. Buenos Aires: Grama Navarin.

Chamorro, J. (2011). *¡Interpretar!* Buenos Aires: Grama.

Delgado, S., Strawn, J., and Pedapati, E. (2015). *Contemporary Psychodynamic Psychotherapy for Children and Adolescents. Integrating Intersubjectivity and Neuroscience*. Berlín: Springer.

Esqué, X. (2017). ¿Por qué analizarse hoy? Conferencia Facultad de Psicología, UNC. Córdoba.

Freud, S. [1905] (2011ll). Fragmento de análisis de un caso de histeria. In *Sigmund Freud. Obras Completas*. Tomo VII. Buenos Aires: Amorrortu.

Freud, S. [1915] (2012g). Pulsiones y destinos de pulsión. In *Sigmund Freud. Obras Completas*. Tomo XIV Buenos Aires: Amorrortu.

García de Frutos, H. (2011). El psicoanálisis no es una ciencia. In *Freudiana* (62). ELP de la EFP miembro de la AMP. Catalunya: Repro Disseny.

Grinbaum, G. (2020). ¿Para qué sirve el psicoanalista? Ciclo conservatorio promovido por Grupo Lacaniano de Formosa.

Han, B.-C. (2022). *Capitalismo y pulsión de muerte*. Barcelona: Herder.

Jones, E. (1981). *Vida y obra de Sigmund Freud*. Tomo 1. Barcelona: Anagrama.

Kandel, E. (2009a). Aspiraciones de la biología para un nuevo humanismo. In E. Kandel (ed.), *Psiquiatría, psicoanálisis, y la nueva biología de la mente*. Tercera ed. España, Barcelona: Ars Medica.

Kandel, E. (2009b). La influencia del pensamiento psiquiátrico en la investigación neurobiológica. In E. Kandel (ed.), *Psiquiatría, psicoanálisis, y la nueva biología de la mente*. Tercera ed. España, Barcelona: Ars Medica.

Lacan, J. [1974] (2006b). *El triunfo de la religión*. Buenos Aires: Paidós.

Lacan, J. [1972–1973] (2008f). *El Seminario. Libro 20. Aun*. Buenos Aires: Paidós.

Lacan, J. [1965] (2012h). Los cuatro conceptos fundamentales del psicoanálisis. In *Otros escritos*. Buenos Aires: Paidós.

Lacan, J. [1964] (2013). *El Seminario. Libro 11. Los cuatro conceptos fundamentales del psicoanálisis*. Buenos Aires: Paidós.

Laurent, E. (2004). Principios rectores del acto analítico. Publicación de EOL. Accessed 13 July 2020 from: www.eol.org.ar/template.asp?Sec=publicaciones&SubSec=on_line&File=on_line/laurent/documentos.html

Laurent, E. (2005). *Lost in cognition. El lugar de la pérdida en la cognición*. Buenos Aires: Diva.

Laurent, E. (2011). La ilusión del cientificismo, la angustia de los sabios. In *Freudiana* (62). ELP de la EFP miembro de la AMP. Catalunya: Repro Disseny.

Laurent, E. (2014b). El psicoanálisis no es una psicoterapia, pero […] In *Freudiana* (70). ELP de la EFP miembro de la AMP. Catalunya: Repro Disseny.

Mías, C. (2008). *Principios de neuropsicología clínica con orientación ecológica. Aspectos teóricos y procedimentales*. Córdoba: Encuentro.

Miller, J.-A. (2016a). Habeas corpus. Intervención pronunciada en la clausura del X congreso de la Asociación Mundial de Psicoanálisis, "El cuerpo hablante. Sobre el inconsciente en el siglo XXI", Río de Janeiro, 25–28 April 2016. En esta secuencia titulada "De Río a Barcelona" intervinieron también Miquel Bassols and Guy Briole.

Miller, J.-A. (2019a). Freud por delante de Lacan. In *Freudiana* (86) "Inconsciente y cerebro: nada en común". ELP de la EFP miembro de la AMP. Catalunya: Repro Disseny.

Miller, J.-A. [1987–1988] (2019b). *Causa y consentimiento*. Buenos Aires: Paidós.

Miller, J.-A. (2021b). Dócil a lo trans. Accessed 29 June 2021 from: https://psicoanalisislacaniano.com/2021/04/22/jam-docil-al-trans-20210422/

Miller, J.-A. [1984–1985] (2021d). *1, 2, 3, 4*. Buenos Aires: Paidós.

Popper, K. (2001). *Conocimiento objetivo. Un enfoque evolucionista*. 4th ed. Madrid: Tecnos.

Regnault, F. (2004). La prueba en psicoanálisis. In *Revista Lacaniana. Las prácticas de la escucha y sus argumentos* (2). Buenos Aires: EOL.

Ritvo, J. (2014). *La retórica conjetural o el nacimiento del sujeto*. Rosario: Nube Negra.

Solms, M. (2006). Neuropsicoanálisis. Entrevista por Steve Ayan. In *Mente y cerebro. Freud. Investigación y ciencia* (18), 74.

Final thoughts

With such a wide field it has been impossible to acquire an exhaustive knowledge of all the literature. Surely there are many important works that were not considered. We declare ourselves guilty of not having made a tour of all the aspects, theoretical perspectives, and schools of psychoanalysis that existed from the time of Freud to the present day that would have allowed us to fully show the state of the art. But where the work of thinking and writing necessarily absorbs so much time and the work of writing demands, then, a logical reduction, such ignorance, although lamentable, does not seem to be completely unforgivable.

In line with the opening movements of the unconscious, this investigation is not closed. We have simply dealt with elementary things of psychoanalysis. However, we believe that there is something that at least makes this text innovative, a new angle on the subject. While other books on psychoanalysis refer to the neurosciences without exhaustively covering them, this book has investigated, like no other book on psychoanalysis, the publications of prominent neuroscientists worldwide to understand how they think and extract the conceptions of the unconscious that they hold.

Thus, Chapter 1 was the key that opened this study. The authors that we read there point to the convergent validity between psychoanalysis and neuroscience, to a use of psychoanalysis based on making it scientific that results in the superimposition of neuroscience on psychoanalysis. Along this path, we find that great researchers direct their efforts to sustain that Freud would have been delighted with this combination. So, in the second chapter, we went back to Freud and showed that he would have been disappointed in this combination. For 52 years Freud maintained that psychoanalysis has nothing to do with neurobiology. To say that the future of psychoanalysis depends on biology is to reject Freud's claim.

With the reading of Lacan, we were able to demonstrate that the positions tending toward the integration of the unconscious in neurobiology would make the variants of psychoanalytic praxis refer to an adaptation of the cure on the basis of empirical, classificatory, similar, and normative criteria, which it would mean the abolition of psychoanalysis instead of its "salvation", which is what they proclaim. Lacan (2012g) warned that psychoanalysis was being diverted while its progress was dampened by degrading its use. His intention, when founding his school, was

DOI: 10.4324/9781003458470-7

to return to lead the original praxis that Freud instituted under the name of psychoa-nalysis to the duty that corresponds to it in our world. This text was motivated by these deviations and found that the discussion is not so much about the differences between psychoanalysis and neuroscience, but about what is psychoanalysis and what is not.

We recognize that this book can be judged for its lack of self-criticism, and for the risk of falling into assumptions and estimates with strong epistemic biases. Per-haps the development of the theme has been reduced to a tension that we all know: Lacanian psychoanalysis versus the neurosciences. Perhaps the reader may ques-tion that we have not gone beyond the formulation of the basic problems. To do this we would have to leave the theoretical identifications of "I am a Lacanian" or "I am a neuropsychoanalyst". That is, to problematize this tension, one must completely get rid of identifications. We cannot claim to have achieved this.

To finish, we want to leave a thoughtful message. In this age that tends to devalue knowing the other, that favors self-management, the "self-taught", that pushes to the liquidity of love ties, that removes the Other, that promotes self-perception, self-designation, psychoanalysis proposes the transference as healing and that im-plies introducing time as an unavoidable variable. Psychoanalysis does not propose solutions in record time, but a listening not in favor of a solution for all. It is about giving the subject a space, a time to work, not brief therapies to resolve everything immediately, but a stopping point toward elaboration, where the subject, taken by excess and speed, can find, in each session, his own solution beginning with a work of knowing. That is our proposal as practitioners of psychoanalysis today.

Reference

Lacan, J. [1965] (2012g). Acto de fundación. In *Otros escritos*. Buenos Aires: Paidós.

Bibliography

Acosta, S. (2020). *Die Traumdeutung*, entre el deseo de dormir y el despertar. In *Scilicet El sueño. Su interpretación y su uso en la cura lacaniana. Publicación en razón del XII congreso de la Asociación Mundial de Psicoanálisis.* Buenos Aires: Grama.

Aflalo, A. (2011). *El intento de asesinato del psicoanálisis.* Buenos Aires: Grama.

Aguiar, A. (2018). The "real without law" in psychoanalysis and neurosciences. *Frontiers in Psychology*, 9: 851. doi: 10.3389/fpsyg.2018.00851

Ansermet, F. and Magistretti, P. (2002). Introducción. In Nathalie Georges, Nathalie Marchaison, and Jacques-Alain Miller, *De las neurociencias a las logociencias, ¿Quiénes son sus psicoanalistas?* Paris: du Seuil.

Ansermet, F. and Magistretti, P. (2006). *A cada cual su cerebro. Plasticidad neuronal e inconsciente.* Buenos Aires: Katz.

Ansermet, F. and Magistretti, P. (2010). L' "île" de la pulsion. In *Les énigmes du plaisir.* Paris: Odile Jacob.

Ansermet, F. and Magistretti, P. (2012). The island of the drive. *Swiss Archives of Neurology and Psychiatry*, 163 (8), 281–285.

Arenas, G. (2018). Estructura lógica de la interpretación. Olivos: Grama.

Arendt, H. (2003). *Eichmann en Jerusalén. Un estudio sobre la banalidad del mal.* Barcelona: Lumen.

Ayan, S. (2006). Mecanismos del inconsciente. In *Mente y cerebro. Freud. Investigación y ciencia* (18), 62–67.

Balzarini, M. (2021). La formación en psicoanálisis de orientación lacaniana y en neurociencias psicoanalíticas. In *Escritos de Posgrado*, año 1, N° 3. ISBN 9789877024036 en línea: Facultad de Psicología, Universidad Nacional de Rosario. Accessed 23 September 2021 from: https://escritosdeposgrado-fpsico.unr.edu.ar/?p=377

Balzarini, M. (2023a). ¿Prevención del acto suicida? en Revista Psicoanálisis en la universidad N°7. Rosario, Argentina, UNR Editora, pp. 145–153. Accessed 2 June 2023 from: https://psicoanalisisenlauniversidad.unr.edu.ar/index.php/RPU/article/view/ 155/118

Balzarini, M. (2023b). Lo inconsciente en psicoanálisis. Un estudio preliminar. México: El Diván Negro.

Bandler, R. and Grinder, J. (1980). *La estructura de la magia. Lenguaje y terapia.* Santiago de Chile: Cuatrovientos.

Barros, M. (2004). La salud de los nominalistas. Un estudio sobre las prácticas psicoterapéuticas. In *Revista Lacaniana. Las prácticas de la escucha y sus argumentos* (2). Buenos Aires: EOL.

Bassols, M. (2011a). *Tu yo no es tuyo*. Buenos Aires: Tres Haches.

Bassols, M. (2011b). Las neurociencias y el sujeto del inconsciente. Conferencia pronunciada en Granada. Instituto del Campo Freudiano. Accessed from: http://www.icf-granada.net/2012-04-04-08-33-03/videos/83-las-neurociencias-y-el-sujeto-del-inconsciente

Bassols, M. (2012a). Lo real del psicoanálisis. Lo real en la ciencia y el psicoanálisis. Virtualia (25). Accessed 17 July 2020 from: www.revistavirtualia.com/articulos/262/lo-real-en-la-ciencia-y-el-psicoanalisis/lo-real-del-psicoanalisis

Bassols, M. (2012b). Psicoanálisis, sujeto y neurociencias. Presentación del libro *Sutilezas analíticas* en Alianza Francesa de San Ángel. Nueva Escuela Lacaniana. Mexico D.F.

Bassols, M. (2012c). Lo real del psicoanálisis. In *Freudiana* (64). ELP de la EFP miembro de la AMP. Catalunya: Repro Disseny.

Bassols, M. (2013a). La vigencia del psicoanálisis. Entrevista por Bordon, J.M. in Revista *Noticias* (pp. 118–120). Centro de investigación y docencia en psicoanálisis. Lima. Accessed from: www.enapol.com/images/Prensa/13-12-06_Entrevista-a-Miquel-Bassols.pdf

Bassols, M. (2013b). La diferencia entre psicoanálisis y ciencia en torno a la idea de cuerpo. Entrevista en *EOL Rosario* por Manuel Ramírez. Accessed from: www.eol.org.ar/template.asp?Sec=prensa&SubSec=america&File=america/2013/13-12-01_Entrevista-a-Miquel-Bassols.html

Bassols, M. (2014). El ocaso de la psiquiatría, ¿y después? In *Freudiana* (72). ELP de la EFP miembro de la AMP. Catalunya: Repro Disseny.

Bassols, M. (2016a). Freud era un misógino contrariado, pero se dejó enseñar por las mujeres. Entrevista en *El País* por Ángela Molina. Accessed from: www.eol.org.ar/template.asp?Sec=prensa&SubSec=europa&File=europa/2016/16-03-27_Entrevista-a-Miquel-Bassols.html

Bassols, M. (2016b). La fascinación mecánica. In *Freudiana* (77–78). ELP de la EFP miembro de la AMP. Catalunya: Repro Disseny.

Bassols, M. (2016c). La "substancia gozante". In *Revista Lacaniana* (21). El racismo que me habita. Buenos Aires: Escuela de la Orientación Lacaniana.

Bassols, M. (2019). El falso dualismo entre mente y cuerpo. In *Freudiana* (86) "Inconsciente y cerebro: nada en común". ELP de la EFP miembro de la AMP. Catalunya: Repro Disseny.

Bassols, M., Laurent, E., and Berenguer, E. (2006). Lost in cognition. In *Freudiana* (46). ELP de la EFP miembro de la AMP. Catalunya: Repro Disseny.

Baudelaire, C. (1856). Las flores del mal. Poesía. Piezas condenadas. Accessed from 18 November 2020: www.cjpb.org.uy/wp-content/uploads/repositorio/serviciosAlAfiliado/librosDigitales/Baudelaire-Flores-Mal.pdf

Bazan, A. (2006). Primary process language. *Neuro-Psychoanalysis*, 8, 157–159.

Bazan, A. (2011). Phantoms in the Voice: A Neuropsychoanalytic hypothesis on the structure of the unconscious. *Neuropsychoanalysis*, 13 (2), 161–176.

Bazan, A. (2012). From sensorimotor inhibition to Freudian repression: insights from psychosis applied to neurosis. *Front. Psychol.*, 3, 452.

Bazan, A. and Detandt, S. (2013). On the physiology of jouissance: interpreting the mesolimbic dopaminergic reward functions from a psychoanalytic perspective. *Frontiers in Human Neuroscience*, 7(709). doi:10.3389/fnhum.2013.00709.

Bazan, A., Detandt, S., and Van de Vijver, G. (2017). The mark, the thing, and the object: On what commands repetition in Freud and Lacan. *Frontiers in Psychology*, 8 (22). doi: 10.3389/fpsyg.2017.02244.

Bazan, A., Shevrin, H., Brakel, L., and Snodgrass, M. (2007). Motivations and emotions contribute to a-rational unconscious dynamics: evidence and conceptual clarification. *Cortex*, 43 (8), 1104.

Bazan, A. and Snodgrass, M. (2012). On unconscious inhibition: instantiating repression in the brain. In A. Fotopoulou, D. W. Pfaff, and E. M. Conway (eds.), *Trends in Psychodynamic Neuroscience*. Oxford: Oxford University Press, 307–337.

Bazan, A., Van Draege, K., De Kock, L., Brakel, L., Geerardyn, F., and Shevrin, H. (2011). Empirical evidence for Freud's theory of primary process mentation in acute psychosis. *Psychoanalytic Psychology*. Advance online publication. Doi: 10.1037/a0027139

Bazan, A. and Zehetner, S. (2018). When people recount their dreams, they don't talk about the hippocampus. In The Vienna Psychoanalyst. Entrevista. Accessed 17 April 2023 from: www.theviennapsychoanalyst.at/index.php?start=2&wbkat=8&wbid=1112

Beck, A., Rush, J. Shaw, B., and Emery, G. (1983). *Terapia cognitiva de la depresión*. Bilbao: Desclee de Brouwer.

Belaga, G. (2011). La salud mental, lo inevitable de una totalidad fallida. In *Lacaniana de psicoanálisis* (11), año VII. Pp. 41–44. Publicación de Escuela de Orientación Lacaniana. Buenos Aires: Grama.

Berridge, K. and Kringelbach, M. (2008). Affective neuroscience of pleasure: reward in humans and animals. *Psychopharmacology*, 199, 457–480. doi: 10.1007/s00213-008-1099-6

Binder, J. and Desai, R. (2011). The neurobiology of semantic memory. *Trends in Cognitive Sciences*, 15 (11), 527–536.

Blanco, M. (2011). La salud mental a la luz de los cuatro conceptos fundamentales del psicoanálisis. In *Freudiana* (61) "Sueño". ELP de la EFP miembro de la AMP. Catalunya: Repro Disseny.

Blass, R. and Carmeli, Z. (2007). The case against neuropsychoanalysis: On fallacies underlying psychoanalysis latest scientific trend and its negative impact on psychoanalytic discourse. *International Journal of Psychoanalysis*, 88, 19–40.

Born, J. and Wagner, U. (2006). ¿Sueñan las redes neuronales?. In *Mente y cerebro. Freud. Investigación y ciencia* (18), 68.

Brakel, L. and Shevrin, H. (2005). Anxiety, attributional thinking, and the primary process. *International Journal of Psycho-Analysis*, 86 (6), 1679–1693.

Briole, G., Rouse, H., Fernández, M., Galaman, C., Godínez, R., and Teixidó, A. (2019). Argumento tercera reunión "La clínica bajo transferencia: nada que ver con los tratamientos neuro". Prólogo al 5° Congreso de la EuroFederación de Psicoanálisis "PIPOL9". Inconsciente y cerebro: nada en común.

Brodsky, G. (2013). Contra la ilusión religiosa. La unión solidaria del psicoanálisis y la ciencia según Freud. Publicación de Escuela de Orientación Lacaniana. Accessed from: www.eol.org.ar/template.asp?Sec=prensa&SubSec=america&File=america/2013/13-08-22_Contra-la-ilusion-religiosa.html

Brodsky, G. (2015a). Mi cuerpo y yo. Conferencia pública. Participan Universidad del Claustro de Sor Juana, local de la NEL-México DF y Alianza Francesa de San Ángel. México. Accessed 9 November 2019 from: www.radiolacan.com/es/topic/589/8#.XU-Q10i5E2oU.whatsapp

Brodsky, G. (2015b). Seminario clínico: "La dirección de la cura". In *Resonancias II*. Revista de Psicoanálisis. Publicación del IOM2 Nuevo Cuyo. Buenos Aires: Grama.

Buchheim, A., Cierpka, M., Kächele, H., and Roth, G. (2013). Efectos del psicoanálisis en el cerebro. In *Mente y cerebro. El legado de Freud. La neurociencia demuestra la eficacia del psicoanálisis. Investigación y ciencia* (62), 26–29.

Bush, G. (2022). Discurso de Bush sobre la importancia de elecciones justas. Dallas, Texas, EStados Unidos. Accessed 27 July 2022 from: www.bbc.com/mundo/media-61509370.amp

Camerer, C., Loewenstein, G., and Prelec, D. (2005). Neuroeconomics: how neuroscience can inform economics. *Journal of Economic Literature,* vol. XLIII, 2005, pp. 9–64.

Cancina, P. (2008). *La investigación en psicoanálisis.* Argentina: Homo sapiens.

Castanet, H. (2023). *Neurología versus psicoanálisis.* Buenos Aires: Grama Navarin.

Chamorro, J. (2011). *¡Interpretar!* Buenos Aires: Grama.

Copjec, J. (2015). *Read My Desire: Lacan Against the Historicists* (2nd ed.). London: Verso Press.

Cosenza, D. (2020). El exceso en el cuerpo del hablanteser. Declinaciones y derivas en la clínica contemporánea. Conferencia. Accessed 16 September 2020 from: www.eol.org.ar/agenda/evento_escuela.asp?Evento=976/Conferencia-de-Domenico-Consenza

Cosenza, D. and Puig, S. (2019). La tentación neurobiológica del psicoanálisis y el corte de Lacan. Presentación Hacia Pipol 9: El inconsciente y el cerebro: nada en común. Escuela Lacaniana de Psicoanálisis de Catalunya del Campo Freudiano. Accessed 15 November 2020 from: www.cdcelp.org/es/ficha-actividad.php?f=396&s=1

Cuñat, C. (2019). El paradigma neurocientífico y el imaginario social. In *Freudiana* (86) "Inconsciente y cerebro: nada en común". ELP de la EFP miembro de la AMP. Catalunya: Repro Disseny.

Dall'Aglio, J. (2019). Of brains and Borromean knots: A Lacanian meta-neuropsychology. *Neuropsychoanalysis,* 21 (1), 23–38. DOI: 10.1080/15294145.2019.1619091

Dall'Aglio, J. (2020a). No-Thing in common between the unconscious and the brain: on the (im)possibility of Lacanian Neuropsychoanalysys. *ResearchGate, Psychoanalysis Lacan,* 4. Accessed 15 April 2023 from: https://researchgate.net/publication/342870600

Dall'Aglio, J. (2020b). Sex and prediction error, part 2: *jouissance* and the free energy principle in neuropsychoanalysis. *Japa,* 69 (4), 715–741. DOI: 10.1177/00030651211042377

Dall'Aglio, J. (2021). What can psychoanalysis learn from neuroscience? A theoretical basis for the emergence of a neuropsychoanalytic model. *Contemporary Psychoanalysis,* 57 (1), 125–145, DOI: 10.1080/00107530.2021.1894542

Damásio, A. (1994). *El error de Descartes. La razón de las emociones.* Buenos Aires: Andres Bello.

Davidovich, M. and Winograd, M. (2010). Psicoanálisis y neurociencias: un mapa de los debates. *Psicologia em Estudo* (15), n. 4, 801–809.

De Georges, P. (2005). Paradigma de desencadenamiento. In Jacques-Alain Miller et al., *Los inclasificables de la clínica psicoanalítica.* Buenos Aires : Paidós, pp. 41–46.

Decety, J. (1996). Neural representations for action. *Rev. Neuroscience,* 7, 285–297

Dehaene, S. (2015). *La conciencia en el cerebro. Descifrando el enigma de cómo el cerebro elabora nuestros pensamientos.* Buenos Aires: Sigloveintiuno.

Delgado, O. (2018). *Huellas freudianas en la última enseñanza de Lacan. Volumen III. La clínica de lo real en Freud.* Buenos Aires: Grama.

Delgado, O. (2021). *Leyendo a Freud desde un diván lacaniano.* Buenos Aires: Grama.

Delgado, S., Strawn, J., and Pedapati, E. (2015). *Contemporary Psychodynamic Psychotherapy for Children and Adolescents.* Integrating Intersubjectivity and Neuroscience. Berlin: Springer.

Deneke, F.-W. (2006). Un modelo estructural revisado. In *Mente y cerebro. Freud. Investigación y ciencia* (18), 71.

Dessal, G. (2016). Vicisitudes actuales de la vida amorosa. Conferencia. Accessed 25 May 2023 from: www.youtube.com/watch?v=nGUxYJVc58U

Edelman, G. and Tononi, G. (2002). *El universo de la conciencia. Cómo la materia se convierte en imaginación.* Barcelona: Crítica.

Eichenbaum, H., Cahill, L., Gluck, M., Hasselmo, M., Keil, F., Martin, A., and Williams, C. (1999). Learning and memory: Systems analysis. In M. Zigmond, F. Bloom, S. Landis, J. Roberts, and L. Squire (eds.), *Fundamental Neuroscience.* New York: Academic Press, pp. 1455–1486.

Ellis, A. (2000). *Vivir en una sociedad irracional: Una guía para el bienestar mediante la terapia racional-emotivo-conductual.* Barcelona: Paidós.

Esqué, X. (2017). ¿Por qué analizarse hoy? Conferencia Facultad de Psicología, UNC. Córdoba.

Fajnwaks, F. (2019). De la industria de la imagen a los usos del soñar. In *Freudiana* (86) Inconsciente y cerebro. Escuela Lacaniana de Psicoanálisis de Catalunya.

Fernández, A. (2019). Discurso en Plaza de Mayo. Accessed 27 July 2022 from: www.lanacion.com.ar/politica/volvimos-y-vamos-a-ser-mujeres-furcio-o-guino-de-alberto-fernandez-nid2314567/

Freud, S. [1895] (1986a). *Sigmund Freud Cartas a Wilhlelm Flies (1887–1904).* Buenos Aires: Amorrortu.

Freud, S. [1916] (1986b). Una dificultad del psicoanálisis. In *Sigmund Freud. Obras Completas.* Tomo XVII. Buenos Aires: Amorrortu.

Freud, S. [1926] (2005). El valor de la vida. Entrevista a Sigmund Freud realizada por George Silvestre Viereck. Traducida del inglés al portugués por Paulo César Souza y al castellano por Miguel Ángel Arce. In *La Brújula* (28), Semanario de la Comunidad Madrileña de la ELP. Directora: Marta Davidovich.

Freud, S. [1925] (2006a). Presentación autobiográfica. In *Sigmund Freud. Obras Completas.* Tomo XX. Buenos Aires: Amorrortu.

Freud, S. [1926] (2006b). ¿Pueden los legos ejercer el análisis? En *Sigmund Freud. Obras Completas.* Tomo XX. Buenos Aires: Amorrortu.

Freud, S. [1935] (2006c). La sutileza de un acto fallido. In *Sigmund Freud. Obras Completas.* Tomo XXII. Buenos Aires: Amorrortu.

Freud, S. [1932] (2006d). 31ª conferencia. La descomposición de la personalidad psíquica. In *Sigmund Freud. Obras Completas.* Tomo XXII. Buenos Aires: Amorrortu.

Freud, S. [1932] (2006e). 34ª conferencia. Esclarecimientos, aplicaciones, orientaciones. In *Sigmund Freud. Obras Completas.* Tomo XXII. Buenos Aires: Amorrortu.

Freud, S. [1932] (2006f). 35ª conferencia. In torno de una cosmovisión. In *Sigmund Freud. Obras Completas.* Tomo XXII. Buenos Aires: Amorrortu.

Freud, S. [1940] (2006g). Esquema del psicoanálisis. In *Sigmund Freud. Obras Completas.* Tomo XXIII. Buenos Aires: Amorrortu.

Freud, S. [1923] (2007b). El yo y el ello. In *Sigmund Freud. Obras Completas.* Tomo XIX. Buenos Aires: Amorrortu.

Freud, S. [1924] (2007c). Breve informe sobre el psicoanálisis. In *Sigmund Freud Obras Completas.* Tomo XIX. Buenos Aires: Amorrortu.

Freud, S. [1888] (2011a). Histeria. In *Obras Completas.* Tomo I. Buenos Aires: Amorrortu.

Freud, S. [1930] (2011b). El malestar en la cultura. In *Sigmund Freud. Obras Completas.* Tomo XXI. Buenos Aires: Amorrortu.

Freud, S. [1927] (2011c). El porvenir de una ilusión. In *Sigmund Freud. Obras Completas.* Tomo XXI. Buenos Aires: Amorrortu.

Freud, S. [1905] (2011d). Tres ensayos de teoría sexual. In *Sigmund Freud. Obras Completas.* Tomo VII. Buenos Aires: Amorrortu.

Freud, S. [1913] (2011e). El interés por el psicoanálisis. In *Sigmund Freud. Obras Completas*. Tomo XIII. Buenos Aires: Amorrortu.

Freud, S. [1886] (2011f). Informe sobre mis estudios en París y Berlín. Realizados con una beca de viaje del Fondo de Jubileo de la Universidad (October 1885–March 1886). In *Sigmund Freud. Obras Completas*. Tomo I. Buenos Aires, Argentina: Amorrortu.

Freud, S. [1896] (2011g). Carta 52. In *Sigmund Freud. Obras Completas*. Tomo I. Buenos Aires: Amorrortu.

Freud, S. [1897] (2011h). Carta 69. In *Sigmund Freud. Obras Completas*. Tomo I. Buenos Aires: Amorrortu.

Freud, S. [1890] (2011i). Tratamiento psíquico (tratamiento del alma). In *Sigmund Freud. Obras Completas*. Tomo I. Buenos Aires: Amorrortu.

Freud, S. [1893] (2011j). Algunas consideraciones con miras a un estudio comparativo de las parálisis motrices orgánicas e histéricas. In *Sigmund Freud. Obras Completas*. Tomo I. Buenos Aires: Amorrortu.

Freud, S. [1895] (2011k). Proyecto de una psicología para neurólogos. In *Sigmund Freud. Obras Completas*. Tomo I. Buenos Aires: Amorrortu.

Freud, S. [1916] (2011l). 18ª conferencia. La fijación al trauma, lo inconciente. In *Sigmund Freud. Obras Completas*. Tomo XVI. Buenos Aires: Amorrortu.

Freud, S. [1905] (2011ll). Fragmento de análisis de un caso de histeria. In *Sigmund Freud. Obras Completas*. Tomo VII. Buenos Aires: Amorrortu.

Freud, S. [1915] (2011m). 2ª conferencia. Los actos fallidos. In *Sigmund Freud. Obras Completas*. Tomo XV. Buenos Aires: Amorrortu.

Freud, S. [1896] (2011n). Manuscrito K. Las neurosis de defensa. In *Obras Completas*. Tomo I. Buenos Aires: Amorrortu.

Freud, S. [1917] (2011ñ). 27ª conferencia. La trasferencia. In *Sigmund Freud. Obras Completas*. Tomo XVI. Buenos Aires: Amorrortu.

Freud, S. [1892] (2011o). Prólogo y notas de la traducción de J.-M. Charcot, *Leçons du mardi de la Salpêtrière*. In *Sigmund Freud. Obras Completas*. Tomo I. Buenos Aires, Argentina: Amorrortu.

Freud, S. [1896] (2011p). Carta 46. Fragmentos de la correspondencia con Fliess. In *Obras Completas*. Tomo I. Buenos Aires: Amorrortu.

Freud, S. [1910] (2012a). La perturbación psicógena de la visión según el psicoanálisis. In *Obras Completas*. Tomo XI. Buenos Aires: Amorrortu.

Freud, S. [1915] (2012b). Lo inconsciente. In *Sigmund Freud. Obras Completas*. Tomo XIV Buenos Aires: Amorrortu.

Freud, S. [1900] (2012c). La interpretación de los sueños. Cap. VII. In *Sigmund Freud. Obras Completas*. Tomo V. Buenos Aires: Amorrortu.

Freud, S. [1901] (2012d). Psicopatología de la vida cotidiana. Cap. I: El olvido de los nombres propios. In *Sigmund Freud. Obras Completas*. Tomo VI. Buenos Aires: Amorrortu.

Freud, S. [1912] (2012e). Consejos al médico sobre el tratamiento psicoanalítico. In *Obras Completas*. Tomo XII. Buenos Aires: Amorrortu.

Freud, S. [1911] (2012f). Sobre psicoanálisis. In *Obras Completas*. Tomo XII. Buenos Aires: Amorrortu.

Freud, S. [1915] (2012g). Pulsiones y destinos de pulsión. In *Sigmund Freud. Obras Completas*. Tomo XIV Buenos Aires: Amorrortu.

Freud, S. [1920] (2012h). Más allá del principio de placer. In *Sigmund Freud. Obras Completas*. Tomo XVIII. Buenos Aires: Amorrortu.

Freud, S. [1912] (2012i). Nota sobre el concepto de lo inconsciente en psicoanálisis. In *Obras Completas*. Tomo XII. Buenos Aires: Amorrortu.

Freud, S. [1913] (2012k). Sobre la iniciación del tratamiento. In *Obras Completas*. Tomo XII. Buenos Aires: Amorrortu.

Freud, S. [1914] (2012ll). Introducción del narcisismo. In *Sigmund Freud. Obras Completas*. Tomo XIV Buenos Aires: Amorrortu.

Fridman, P. and Millas, D. (2005a). La exaltación maníaca. Las muertes del sujeto. In Jacques-Alain Miller et al., *Los inclasificables de la clínica psicoanalítica*. Buenos Aires: Paidós, pp. 81–87

Fridman, P. and Millas, D. (2005b). Segunda discusión. La muerte del sujeto. In Jacques-Alain Miller et al., *Los inclasificables de la clínica psicoanalítica*. Buenos Aires: Paidós, pp. 89–98.

Gallese, V. (2000). The inner sense of action: agency and motor representations. *J. Conscious. Stud.*, 7, 23–40

García de Frutos, H. (2011). El psicoanálisis no es una ciencia. In *Freudiana* (62). ELP de la EFP miembro de la AMP. Catalunya: Repro Disseny.

García de Frutos, H. (2012). Neurocientificismo, logicismo y psicoanálisis: algunos apuntes para una perspectiva crítica. In *Freudiana* (65) "Los espectros del autismo". ELP de la EFP miembro de la AMP. Catalunya: Repro Disseny.

Goodman, N. (1993). *Hecho, ficción y pronóstico*. Madrid, España: Síntesis.

Goya, A. (2017). *Cinco conferencias sobre psicosis ordinaria*. Olivos, Argentina: Grama.

Grinbaum, G. (2020). ¿Para qué sirve el psicoanalista? Ciclo conservatorio promovido por Grupo Lacaniano de Formosa.

Grinbaum, G. (2022). El hijo adolescente de Harry Potter. In *Rayuela* (9). Accessed 23 November 22 from: www.revistarayuela.com/es/009/template.php?file=notas/de-padres-e-hijos-en-el-mundo-de-la-inexistencia-del-otro.html

Han, B.-C. (2012). *La sociedad del cansancio*. Barcelona: Herder.

Han, B.-C. (2014). *La agonía del Eros*. Barcelona: Herder.

Han, B.-C. (2022). *Capitalismo y pulsión de muerte*. Barcelona: Herder.

Handke, P. (2006). *Ensayo sobre el cansancio*. Madrid: Alianza.

Hegel, G. (1966). *Fenomenología del espíritu*. Mexico D. F.: F.C.E. España S.A. Ed.

Ibáñez, A. (2017). ¿Qué son las neurociencias? Noche de la EOL. In *e-Mariposa* (10). Temas de psiquiatría y psicoanálisis. Revista del Departamento de Estudios sobre Psiquiatría y Psicoanálisis (ICF-CICBA). Buenos Aires: Grama, pp. 25–31.

Insel, T. (2009). Un nuevo marco intelectual para la psiquiatría. In E. Kandel (ed.), *Psiquiatría, psicoanálisis, y la nueva biología de la mente*. Tercera edición. España, Barcelona: Ars Medica.

Johnston, A. (2013). Drive between brain and subject: an immanent critique of Lacanian neuropsychoanalysis. *The Southern Journal of Philosophy*, 51 (Spindel Supplement), 48–84.

Jones, E. (1981). *Vida y obra de Sigmund Freud*. Tomo 1. Barcelona: Anagrama.

Judith, L. and Rapoport, M. (2009). La psicoterapia y la sinapsis única. In E. Kandel (ed.), *Psiquiatría, psicoanálisis, y la nueva biología de la mente*. Tercera edición. España, Barcelona: Ars Medica.

Kafka, F. (2011). *Informe para una Academia*. Buenos Aires: Maldoror.

Kandel, E. (2001a). Psychotherapy and the single synapse: the impact of psychiatric thought on neurobiological research. *Neuropsychiatry Clin, Neuroscience*, 13 (2), 290–300.

Kandel E. (2001b). The molecular Biology of Memory Storage: a dialogue between Genes and Synapses. *Science*, 294, 1030–1038.

Kandel, E. (2007). *En busca de la memoria. El nacimiento de una nueva ciencia de la mente.* Buenos Aires: Katz conocimiento.

Kandel, E. (2009a). Aspiraciones de la biología para un nuevo humanismo. In E. Kandel (ed.), *Psiquiatría, psicoanálisis, y la nueva biología de la mente.* Tercera edición. España, Barcelona: Ars Medica.

Kandel, E. (2009b). La influencia del pensamiento psiquiátrico en la investigación neuro-biológica. In E. Kandel (ed.), *Psiquiatría, psicoanálisis, y la nueva biología de la mente.* Tercera edición. España, Barcelona: Ars Medica.

Kandel, E. (2018). *The Disordered Mind.* Nueva York: Farrah, Strauss and Giroux.

Kandel, E., Schwartz, J., and Jessell, T. (2001) *Principios de neurociencia.* Cuarta ed. España: McGraw Hill Interamericana España.

Kernberg, O. (1998). *Love relations.* Inglaterra: Yale University Press.

Koretzky, C. (2019). Sueños y despertares. Una elucidación lacaniana. Buenos Aires: Grama

Kuhn, T. (1964). *La estructura de las revoluciones científicas.* "Posdata". México: Fondo de cultura económica.

Lacan, J. [1966] (1985). Psicoanálisis y medicina. In *Intervenciones y textos.* Buenos Aires: Manantial.

Lacan, J. [1974] (1988a). La tercera. In *Intervenciones y textos 2.* Buenos Aires: Manantial.

Lacan, J. [1975–1976] (2006a). *El Seminario. Libro 23. El sinthome.* Buenos Aires: Paidós.

Lacan, J. [1974] (2006b). *El triunfo de la religión.* Buenos Aires: Paidós.

Lacan, J. [1969–1970] (2007). *El Seminario. Libro 10. La angustia.* Buenos Aires: Paidós.

Lacan, J. [1954–1955] (2008a). *El Seminario. Libro 2. El yo en la teoría de Freud y en la técnica psicoanalítica.* Buenos Aires: Paidós.

Lacan, J. [1969–1970] (2008e). *El Seminario. Libro 17. El Reverso del Psicoanálisis.* Buenos Aires: Paidós.

Lacan, J. [1972–1973] (2008f). *El Seminario. Libro 20. Aun.* Buenos Aires: Paidós.

Lacan, J. [1965] (2009a). La ciencia y la verdad. In *Escritos 2.* Buenos Aires: Sigloveintiuno.

Lacan, J. [1970–1971] (2009b). *El Seminario. Libro 18. De un discurso que no fuera del semblante.* Buenos Aires: Paidós.

Lacan, J. [1955] (2009c). La cosa freudiana o sentido del retorno a Freud en psicoanálisis. In *Escritos 1.* Buenos Aires, Argentina: Sigloveintiuno.

Lacan, J. [1953] (2009d). Función y campo de la palabra y del lenguaje en psicoanálisis. In *Escritos 1.* Buenos Aires, Argentina: Sigloveintiuno.

Lacan, J. [1951] (2009e). Intervención sobre la transferencia. In *Escritos 1.* Bs Aires: Sigloveintiuno.

Lacan, J. [1955–1956] (2009f). *El Seminario. Libro 3. Las psicosis.* Buenos Aires: Paidós.

Lacan, J. [1946] (2009g). Acerca de la causalidad psíquica. In *Escritos 1.* Buenos Aires, Argentina: Sigloveintiuno.

Lacan, J. [1957] (2009h). La instancia de la letra en el inconsciente, o la razón desde Freud. In *Escritos 1.* Buenos Aires: Sigloveintiuno.

Lacan, J. [1957–1958] (2010). El Seminario. Libro 5. *Las formaciones del inconsciente.* Buenos Aires: Paidós.

Lacan, J. [1971–1972] (2012a). *Hablo a las Paredes.* Buenos Aires: Paidós.

Lacan, J. [1953–1954] (2012b). *El Seminario. Libro 1. Los escritos técnicos de Freud.* Buenos Aires: Paidós.

Lacan, J. [1971–1972] (2012c). *El Seminario. Libro 19 [...]o peor.* Buenos Aires: Paidós.

Lacan, J. [1969] (2012d). El acto psicoanalítico. In *Otros escritos*. Buenos Aires: Paidós.

Lacan, J. [1970] (2012e). Radiofonía. In *Otros escritos*. Buenos Aires: Paidós.

Lacan, J. [1973] (2012f). Televisión. In *Otros escritos*. Buenos Aires: Paidós.

Lacan, J. [1965] (2012g). Acto de fundación. In *Otros escritos*. Buenos Aires: Paidós.

Lacan, J. [1965] (2012h). Los cuatro conceptos fundamentales del psicoanálisis. In *Otros escritos*. Buenos Aires: Paidós.

Lacan, J. [1964] (2013). *El Seminario. Libro 11. Los cuatro conceptos fundamentales del psicoanálisis*. Buenos Aires: Paidos.

Langaney, A. (2006). El sentido de la seducción. In *Mente y cerebro. Freud. Investigación y ciencia* (18), 80–82.

Lardjane, R. (2019). El inconsciente y el cerebro en psiquiatría. In *Lacan cotidiano. Para Pipol 9*. Revista de Psicoanálisis (824). BOLC.

Laurent, E. (1991). Psicoanálisis y ciencia: El vacío del sujeto y el exceso de objetos. In *Freudiana* (3). ELP de la EFP miembro de la AMP. Catalunya: Repro Disseny.

Laurent, E. (2002). *Síntoma y nominación*. Buenos Aires: Diva.

Laurent, E (2004). Principios rectores del acto analítico. Publicación de EOL. Accessed 13 July 2020 from: www.eol.org.ar/template.asp?Sec=publicaciones&SubSec=on_line&File=on_line/laurent/documentos.html

Laurent, E. (2005). *Lost in cognition. El lugar de la pérdida en la cognición*. Buenos Aires: Diva.

Laurent, E. (2006). "Principios rectores del acto analítico". In *Mediodicho N° 31*. Córdoba: EOL Sección Córdoba.

Laurent, E. (2007). ¡Es difícil no estar deprimido! Reportaje por Magdalena Ruiz Guiñazu. In EOL Publicaciones. Accessed 11 April 2019 from: www.eol.org.ar/template.asp?Sec=prensa&SubSec=america&File=america/2007/07_12_09_laurent_reportaje.html

Laurent, E. (2008). Usos de las neurociencias para el psicoanálisis. Comunicación en el coloquio organizado en el Collège de France por Pierre Magistretti, con el título "Neurociencias y psicoanálisis". Accessed 23 August 2019 from: www.wapol.org/es/articulos/Template.asp?intTipoPagina=4&intPublicacion=5&intEdicion=43&intIdiomaPublicacion=1&intArticulo=1447&intIdiomaArticulo=1

Laurent, E. (2011). La ilusión del cientificismo, la angustia de los sabios. In *Freudiana* (62). ELP de la EFP miembro de la AMP. Catalunya: Repro Disseny.

Laurent, E. (2014b). El psicoanálisis no es una psicoterapia, pero [...] In *Freudiana* (70). ELP de la EFP miembro de la AMP. Catalunya: Repro Disseny.

Laurent, E. (2016a). *El reverso de la biopolítica*. Buenos Aires: Grama.

Laurent, E. (2016b). El cuerpo hablante: El inconsciente y las marcas de nuestras experiencias de goce. Entrevista por Marcus André Vieira. In *Lacan cotidiano* (576). Accessed from: www.eol.org.ar/biblioteca/lacancotidiano/LC-cero-576.pdf

Laurent, E. (2020a). Hay tantas fes. Revista Lapso (5). Maestría en teoría psicoanalítica lacaniana. UNC. Accessed 28 August 2020 from: http://matpsil.com/revista-lapso/portfolio-items/laurent-video-videoentrevista/

Laurent, E. (2020b). *El nombre y la causa. Conicet y UNC*. Córdoba: IIPsi Instituto de Investigaciones Psicológicas

Lefort, R. (2012). El camino sobre la cresta de la duna. In *Freudiana* (65) "Los espectros del autismo". ELP de la EFP miembro de la AMP. Catalunya: Repro Disseny.

Maino, F. (2020). La vía real del río Aqueronte. In *Scilicet* El sueño. Su interpretación y su uso en la cura lacaniana. Publicación en razón del XII congreso de la Asociación Mundial de Psicoanálisis. Buenos Aires: Grama.

Manes, F. (2019). Seis consejos para cuidar la salud de tu cerebro. Accessed from: https://aprendemosjuntos.elpais.com/especial/la-vida-no-es-la-que-vivimos-sino-como-la-recordamos-para-contarla-facundo-manes/; Accessed from: www.youtube.com/watch?v=3-18pPudCxM&feature=youtu.be

Martinez, M. (2018). Cuando la ciencia vacila. Entrevista realizada por Silvia Salvarezza y Luis Martínez. In *Revista Lacaniana de Psicoanálisis* (24), Ciencia Ficción, EOL, 13. Buenos Aires, Grama, 193–195.

McCabe, D. and Castel, A. (2008). Seeing is believing: The EECT of brain images on judgments of scientific reasoning. *Cognition,* 107, 343–352.

McGilchrist, I. (2009). *The Master and his Emissary: The Divided Brain and the Making of the Western World.* New Haven: Yale University Press.

Mías, C. (2008). *Principios de neuropsicología clínica con orientación ecológica. Aspectos teóricos y procedimentales.* Córdoba: Encuentro.

Miller, J. (2018). Cientismo, ruina de la ciencia. In *Revista Lacaniana de Psicoanálisis* (24), Ciencia Ficción, EOL, 13. Buenos Aires, Grama, pp. 11–13.

Miller, J.-A. (1978). El hombre neuronal. Entrevista de J.-P. Changeux a J.-A. Miller, É. Laurent, J. Bergès y A. Grosrichard. *Ornicar?* (17/18).

Miller, J.-A. (1987a). "Entrevista con Jacques-Alain Miller" por Gonzalez, F. In *Revista A.E.N.* Vol. VII. N° 23.

Miller, J.-A. (1994b). Psicoterapia y psicoanálisis. In *Revista Freudiana* (10). Escuela Europea de Psicoanálisis-Catalunya.

Miller, J.-A. [1995] (1996c). El olvido de la interpretación. In *Entonces: Shhh [...]* Buenos Aires: Minilibros Eolia.

Miller, J.-A. [1996] (1996d). Apología de la sorpresa. In *Entonces: Shhh [...]* Buenos Aires: Minilibros Eolia.

Miller, J.-A. [1986–1987] (1998a). *Los signos del goce.* Buenos Aires: Paidós.

Miller, J.-A. (1998b). La imagen reina. In *Elucidación de Lacan.* Buenos Aires: Paidós.

Miller, J.-A. [1981] (1998d). Psicoanálisis y psiquiatría. In *Elucidación de Lacan. Charlas brasileñas.* Buenos Aires: Paidós.

Miller, J.-A. (2002) *Biología lacaniana y acontecimiento del cuerpo.* Buenos Aires: Colección Diva.

Miller, J.-A. (2004b). Improvisación sobre *Rerum Novarum.* In *Revista Lacaniana. Las prácticas de la escucha y sus argumentos* (2). Buenos Aires: EOL.

Miller, J.-A. (2004c). Verdad, probabilidad estadística, lo real. In *Revista Lacaniana. Las prácticas de la escucha y sus argumentos* (2). Buenos Aires: EOL.

Miller, J.-A. [1987] (2006). *Introducción al método psicoanalítico.* Buenos Aires: Eolia-Paidós.

Miller, J.-A. [1985–1986] (2010). *Extimidad.* Buenos Aires: Paidós.

Miller, J.-A. [1998–1999] (2011a). Paradigmas del goce. In *La experiencia de lo real en la cura psicoanalítica.* Buenos Aires: Paidós.

Miller, J.-A. [1999] (2011c). Biología lacaniana. In *La experiencia de lo real en la cura psicoanalítica.* Buenos Aires: Paidós.

Miller, J.-A. [1988] (2012a). El psicoanálisis, su lugar entre las ciencias. In *Revista consecuencias* (9). Accessed 17 May 2020 from: www.revconsecuencias.com.ar/ediciones/009/template.php?file=arts/Alcances/El-psicoanalisis-su-lugar-entre-las-ciencias.html

Miller, J.-A. [2000–2001] (2013b). *El lugar y el lazo.* Buenos Aires: Paidos.

Miller, J.-A. [2008–2009] (2014a). *Sutilezas analíticas.* Buenos Aires: Paidós.

Miller, J.-A. [2006–2007] (2014b). *El ultimísimo Lacan.* Buenos Aires: Paidós.

Miller, J.-A. (2014c). El inconsciente y el cuerpo hablante. Conferencia pronunciada por Jacques-Alain Miller en la clausura del IX Congreso de la Asociación mundial de psicoanálisis (AMP) presentando el tema del X Congreso en Río de Janeiro. In *Revista Lacaniana* (17). Buenos Aires: Grama.

Miller, J.-A. [2008] (2015a). *Todo el mundo es loco*. Buenos Aires: Paidós.

Miller, J.-A. [1980] (2015b). *Seminarios de Caracas y Bogotá*. Buenos Aires: Paidos.

Miller, J.-A. (2016a). Habeas corpus. Intervención pronunciada en la clausura del X congreso de la Asociación Mundial de Psicoanálisis, "El cuerpo hablante. Sobre el inconsciente en el siglo XXI", Río de Janeiro, 25–28 April 2016. En esta secuencia titulada "De Río a Barcelona" intervinieron también Miquel Bassols y Guy Briole.

Miller, J.-A. (2016b). ¿Ha dicho raro? In *Revista Mediodicho* (42) ¿A qué le tenemos miedo? Publicación de EOL. Córdoba, Argentina.

Miller, J.-A. (2016c). Sobre el discurso de la ciencia. In *Un esfuerzo de poesía*. Buenos Aires: Paidós.

Miller, J.-A. (2016d). Acción lacaniana. In *Un esfuerzo de poesía*. Buenos Aires: Paidós.

Millar, J.-A. (2017). El niño y el saber. In *Los miedos de los niños*. Buenos Aires: Paidós.

Miller, J.-A. (2017b). Punto de capitón. Accessed 2 June 2021 from: https://psicoanalisislacaniano.com/curso-de-jacques-alain-miller-ano-cero-dictado-en-la-escuela-de-la-causa-freudiana-20170624/#_ftnref5

Miller, J.-A. (2018). Jacques Lacan: observaciones sobre su concepto de pasaje al acto. In Bardón C. and Montserrat P. (eds.), *Suicidio, medicamentos y orden público*. Barcelona: Gredos.

Miller J.-A., (2018c). Neuro-, le nouveau réel. In *La Cause du désir* (98).

Miller, J.-A. (2019a). Freud por delante de Lacan. In *Freudiana* (86) "Inconsciente y cerebro: nada en común". ELP de la EFP miembro de la AMP. Catalunya: Repro Disseny.

Miller, J.-A. [1987–1988] (2019b). *Causa y consentimiento*. Buenos Aires: Paidós.

Miller, J.-A. (2021b). Dócil a lo trans. Accessed 29 June 2021 from: https://psicoanalisislacaniano.com/2021/04/22/jam-docil-al-trans-20210422/

Miller, J.-A. [1984–1985] (2021d). *1, 2, 3, 4*. Buenos Aires: Paidós.

Miller, J.-A. [2011] (2021e). *El ser y el uno*. Los cursos psicoanalíticos de Jacques-Alain Miller. Inédito. Buenos Aires: Paidós.

Miller, J.-A. (2023). El nacimiento del campo freudiano. Conversación con jóvenes. Accessed 26 June 2023 from: www.youtube.com/watch?app=desktop&v=gAVcOuaUyYM

Milner, J.-C. (1996). *La obra clara. Lacan, la ciencia y la filosofía*. Buenos Aires: Manantiales.

Moraga, P. (2019). El paradigma forclusivo de las neurociencias. In *Lacan cotidiano. Para Pipol 9*. Revista de Psicoanálisis (824). BOLC.

Olds, J. (1956). Pleasure centers in the brain. *Science Am.*, 105–116.

Olds, J. and Milner, P. (1954). Positive reinforcement produced by electrical stimulation of septal area and other regions of rat brain. *J. Comp. Physiol. Psychol.*, 47, 419–427. doi: 10.1037/h0058775

Ordóñez, P. (2023). Año nuevo: ¿una nueva vida para un nuevo tiempo? En *Comercio y justicia* 84 (24041), periódico de Ciudad Autónoma de Buenos Aires, 2 de enero 2023.

Panksepp, J. (1998). *Affective Neuroscience: The Foundations of Human and Animal Emotions*. New York: Oxford University Press.

Peteiro, J. (2010). *El autoritarismo científico*. Colección Itaca. Málaga, España: Miguel Gomez.

Pina, A. (2008). *Psiquiatría y psicoanálisis en el marco de las neurociencias*. Barcelona: Biblioteca nueva.

Pinker, S. (2001). *Cómo funciona la mente*. Barcelona: Destino.

Pinker, S. (2004). Órganos de computación. Entrevista realizada por Brockman. In *Revista Lacaniana. Las prácticas de la escucha y sus argumentos* (2). Buenos Aires: EOL.

Pino, S. (2018). La utilidad de las ficciones. In *Revista Lacaniana de Psicoanálisis* (24), Ciencia Ficción, EOL, 13. Buenos Aires, Grama, 50–53.

Pommier, G. (2010). *Cómo las neurociencias demuestran el psicoanálisis*. Buenos Aires: Letra viva.

Popper, K. (2001). *Conocimiento objetivo. Un enfoque evolucionista*. 4th ed. Madrid: Tecnos.

Price, C., Moore, C., Humphreys, G., and Wise, R. (1997). Segregating semantic from phonological processes during reading. *Journal of Cognitive Neuroscience*, 9 (6), 727–733.

Provine, R. (2006). El Bostezo. In *Mente y cerebro. Freud. Investigación y ciencia* (18), 17–25.

Pulice, G., Zelis, O., and Manson, F. (2019). *Investigación <> Psicoanálisis. Fundamentos epistémicos y metodológicos. De Sherlock Holmes, Peirce y Dupin a la experiencia freudiana*. México: El diván negro.

Rapcsak, S., Beeson, P., Henry, M., Leyden, A., Kim, E., Rising, K., and Cho, H. (2009). Phonological dyslexia and dysgraphia: Cognitive mechanisms and neural substractes. *Cortex*, 45 (5), 575–591.

Redmond, J. (2015). Debating the subject: Is there a Lacanian neuropsychoanalysis? Psychoanalysis Lacan, 1. Accessed 9 May 2023 from: https://lacancircle.com.au/wp-content/uploads/2020/09/Debating_the_subject.pdf

Regnault, F. (2004). La prueba en psicoanálisis. In *Revista Lacaniana. Las prácticas de la escucha y sus argumentos* (2). Buenos Aires: EOL.

Ritvo, J. (2014). *La retórica conjetural o el nacimiento del sujeto*. Rosario: Nube Negra.

Rodriguez, T. (2013). Furtivos saboteadores de la salud. In *Mente y cerebro. El legado de Freud. La neurociencia demuestra la eficacia del psicoanálisis. Investigación y ciencia* (62), 5.

Rosales, J. (2017). *La valía de la escritura testimonial para la enseñanza psicoanalítica*. Querétaro, México: Fontamara.

Rose, N. (2011). Gobernar la conducta en la época del cerebro. Conferencia dictada en III Coloquio Latinoamericano de Biopolítica, Universidad Pedagógica (UNIPE). Buenos Aires.

Russell, B. [1903] (1977). *Los principios de la matemática*. Translated by Juan Carlos Grimberg. 3th ed. Madrid: Espasa Calpe.

Sanchez, B. (2018). La ciencia, partenaire incompleto. In *Revista Lacaniana de Psicoanálisis* (24), 13. Buenos Aires: EOL, pp. 67–69.

Sanchez R.-J. and Sanchez C.-J. (2004). Manual de psicoterapia cognitiva (fragmento). In *Revista Lacaniana. Las prácticas de la escucha y sus argumentos* (2). Buenos Aires: EOL.

Shevrin, H. (2003). The psychoanalytic theory of drive in the light of recent neuroscience findings and theories. 1st Annual C. Philip Wilson M. D. Memorial Lecture, New York.

Simonet, P. (2019). Claridad hipnótica del cerebro. In *Lacan cotidiano. Para Pipol 9*. Revista de Psicoanálisis (824). BOLC.

Sinatra, E. (2017). *Las entrevistas preliminares y la entrada en análisis*. Cuadernos del ICdeBA. Buenos Aires: Grama.

Slezak, D. (2018). Una App que ayuda a diagnosticar esquizofrenia a través del análisis del discurso de pacientes. Conicet. UBA. Argentina. Accessed from www.conicet.gov.ar/una-app-que-ayuda-a-diagnosticar-esquizofrenia-a-traves-del-analisis-del-discurso-de-pacientes/

Solms, K. and Solms, M. (2005). *Estudios clínicos en neuropsicoanálisis. Introducción a la neuropsicología profunda*. Bogotá: Fondo de cultura económica.

Solms, M. (2004). Psychanalyse et neurosciences. *Pour la science* (324).

Solms, M. (2006). Neuropsicoanálisis. Entrevista por Steve Ayan en *Mente y cerebro. Freud. Investigación y ciencia* (18), 74.

Solms, M. (2007). Sigmund Freud hoy. *Revista Psicoanálisis*, *5*, 115–119.

Solms, M. (2013). The conscious id. *Neuropsychoanalysis*, 15 (1), 5–19. Accessed 16 April 2023 from: https:// doi.org/10.1080/15294145.2013.10773711

Solms, M. (2015). *The Feeling Brain: Selected Papers on Neuropsychoanalysis*. London: Karnac.

Solms, M. (2017a). What is "the unconscious", and where is it located in the brain? A neuropsychoanalytic perspective. *Annals of the New York Academy of Sciences*, 1406 (1), 90–97.

Solms, M. (2017b). "The unconscious" in psychoanalysis and neuroscience: An integrated approach to the cognitive unconscious. In M. Leuzinger-Bohleber, S. Arnold, and M. Solms (eds.), *The Unconscious: A Bridge Between Psychoanalysis and Cognitive Neuroscience* (pp. 16–35). London: Routledge.

Solms, M. (2020). Entrevista en "Recomendaciones neurocientíficas para los profesionales que practican el psicoanálisis". Seminario virtual de la IPA. Londres.

Solms, M. and Gamwell, L. (2006). *From Neurology to Psychoanalysis. Sigmund Freud's Neurological Drawings and Diagrams of the Mind*. New York: Binghamton University.

Solms, M. and Turnbull, O. (2011). ¿Qué es neuropsicoanálisis? In *Revista Neuropsicoanálisis*, 13 (2). Depto. De Psicología. Universidad Cape Town, Sudáfrica, pp. 133–146.

Solms, M. and Turnbull, O. (2002). *The Brain and the Inner World: An Introduction to the neuroscience of Subjective Experience*. New York: Other Press.

Soria, N. (2020). *La inexistencia del nombre del padre*. Buenos Aires, Argentina: del Bucle.

Stagnaro, J. (2009). "Psiquiatría y neurobiología: el arte de curar y la ciencia del cerebro en crisis paradigmática". In *Jacques Lacan y los matemáticos, los lógicos y los científicos*. Buenos Aires: Escuela Freudiana de Buenos Aires.

Stecco, C. (2020). El ombligo del sueño. Un impoético. In *Scilicet* El sueño. Su interpretación y su uso en la cura lacaniana. Publicación en razón del XII congreso de la Asociación Mundial de Psicoanálisis. Buenos Aires: Grama.

Sulloway, F. (1992). *Freud, Biologist of the Mind: Beyond the Psychoanalytic Legend*. Boston, MA: Harvard University Press.

Talvitie, T. (2009). *Freudian Unconscious and Cognitive Neuroscience. From Unconscious Fantasies to Neural Algorithms*. London: Karnac.

Teboul, D. (producer). Copans, R. and Cohen-Solal, A. (directors). (2019). *Sigmund Freud, un judío sin Dios* [cinta cinematográfica]. Francia: ARTE France y WIL-Dart Film Production. Accessed 14 September 2020 from: https://tv.festhome.com/ff/festival-internacional-de-cine-documental-fidba/861/182455

Teixidó, A. (2019a). Des-cerebrados. In Prólogo al 5° Congreso de la EuroFederación de Psicoanálisis "PIPOL9". Inconsciente y cerebro: nada en común. Accessed 30 May 2020 from: www.cdcelp.org/es/ficha-actividad.php?f=396&s=1

Trilling, L. (1981). Introducción. In E. Jones. *Vida y obra de Sigmund Freud* (pp. 5–19). Barcelona: Anagrama.

Ubieto, J. (2019b). El paradigma "neuro" y las paradojas del goce. In *Prólogo* hacia el 5° Congreso de la EuroFederación de Psicoanálisis "PIPOL9". Inconsciente y cerebro: nada en común. Accessed 30 May 2020 from: www.cdcelp.org/es/ficha-actividad.php?f=396&s=1

Ubieto, J. (2019c). El paradigma "neuro" y las paradojas del goce. In *Freudiana* (86) "Inconsciente y cerebro: nada en común". ELP de la EFP miembro de la AMP. Catalunya: Repro Disseny.

Vanheule, S. (2011). Lacan's construction and deconstruction of the double-mirror device. *Frontiers in Psychology*, 2 (209). doi:10.3389/fpsyg.2011.00209.

Vilá, F. (2019a). La ideología de la Ciencia, la muerte y el sexo. In Prólogo hacia el 5° Congreso de la EuroFederación de Psicoanálisis "PIPOL9". Inconsciente y cerebro: nada en común. Accessed 30 May 2020 from: www.cdcelp.org/es/ficha-actividad.php?f=396&s=1

Voos, D. (2013). Búsqueda del trastorno en el inconsciente. In *Mente y cerebro. El legado de Freud*. La neurociencia demuestra la eficacia del psicoanálisis. *Investigación y ciencia* (62), 22–25.

Wallerstein, R. (2004). Introducción a la mesa redonda sobre psicoanálisis y psicoterapia. La relación entre el psicoanálisis y la psicoterapia. Problemas actuales. In *Revista Lacaniana. Las prácticas de la escucha y sus argumentos* (2). Buenos Aires: EOL.

Yellati, N. (2017). Experimentar con humanos o el investigador en su laboratorio. Noche de la EOL. In *e-Mariposa* (10). Temas de psiquiatría y psicoanálisis. Revista del Departamento de Estudios sobre Psiquiatría y Psicoanálisis (ICF-CICBA). Buenos Aires: Grama, pp. 32–35

Yellati, N. (2018). *Lo que el psicoanálisis enseña a las neurociencias*. Buenos Aires: Grama.

Yellati, N. (2021). Lo que el psicoanálisis enseña a las neurociencias. Conferencia dictada por modalidad virtual a través de Yoica AC. Accessed 24 August 2021 from: https://youtu.be/O22TlWW9bLA

Yue, G. and Cole, K. (1992). Strength increases from the motor program: comparison of training with maximal voluntary and imagined muscle contractions. *J. Neurophysiologie*, 67, 114–123

Zack, O. (2007). *Un uso ético de los antidepresivos. Entrevista para El Tribuno*. Salta: EOL.

Zack, O. (2008). Mesa de apertura. *Discurso de apertura de RedAcción* (27). EOL.

Zack, O. (2016). *Vigencia de las neurosis*. Olivos: Grama.

Zlotnik, M. (2018). Ciencia inexacta. In *Revista Lacaniana de Psicoanálisis* (24), Ciencia Ficción, EOL, 13. Buenos Aires: Grama, pp. 54–56.

Index

accumbens 8, 26, 85, 94
accumulation 27, 36; of data 37
accusations 49, 131, 139; *see also*
   imputations
adaptation 11–12, 18, 26, 30, 81, 117, 139,
   142; *see also* homeostasis
amygdala 9–10, 85–86
analysand 8, 13, 17, 80, 84, 109, 119, 133
Ansermet, F. 3–7, 9–11, 41, 48–49
artificial intelligence 36
awareness 51, 54; affective 39

balance 10–11, 27–28, 71, 92–95, 117, 120
Bazan, A. 7, 14–15, 19, 26–30, 48–49
beyond the pleasure principle 41, 92–96,
   133
body: affective consciousness and 39;
   hysterical paralysis and 23–24; and
   image 44; jouissance and 103–104;
   mind unified with 12, 15, 24–25, 34;
   Other 3, 11, 69–70, 76–77, 79, 89–90,
   92, 95, 102, 107–108; rejection of the
   20, 39, 69, 95, 103–105, 110–112, 115,
   134, 137; *see also* unconscious: being
   that speaks; *parlêtre*
Broca, P. 19, 24, 30

capitalism 35–37, 116, 120, 138; *see also*
   accumulation; body: rejection of the;
   Han, B
cause: analytics 75, 79–82, 91–92, 96, 105,
   111, 119–120, 132–136; cerebral 9–11,
   19–24, 26–29, 31, 44–45, 53, 68, 71–72,
   95, 109–112, 114–116, 119; psychic
   21–22, 69–70
Changeux, J. 4, 31, 110
Charcot, J. 22, 67–69
conscience 75, 114; cognitive 40; moral 93

consciousness 6, 10, 13, 15, 17, 34, 72, 75,
   77, 83, 86, 119; affective 7–8, 39–40;
   see also awareness: affective
contingency 9, 28
correspondence 6, 8; with Fliess 73

Dall'Aglio, J. 3, 10, 14–15, 26–27, 29,
   39–40, 43–45, 48, 93, 112, 123
Damásio, A. 12, 88, 109, 111, 123
Darwin, C. 6, 65, 124
Dehaene, S. 4, 7, 24–25
delirium 74
deprivation 135–136
desire 18, 121; analyst's 32, 47, 65–66, 68,
   80, 111, 116–117, 125–126, 131–133,
   136; fulfillment 132; to know 47, 116;
   motor activation and 28–29; sexual 31;
   unconscious 46, 117
discontinuity 21, 87
discourse: hysteria 8, 68, 76; owner
   85, 108–110, 116–119, 122, 139;
   psychoanalyst 96, 104, 115–116, 120,
   123–124, 133, 135, 137–138; scientific
   8, 35, 43, 45, 83, 87, 96, 116, 123–124,
   139; unconscious 20, 32, 37, 75;
   university 82, 112, 134–136, 138; *see
   also* discourse of the master
dopamine: release 95; system 26–29, 40, 71
dreams 14–16, 18, 21, 32, 35–37, 74, 112,
   118, 121, 125–126
drive 70, 89–90, 113, 121; death 37–38,
   75, 92, 94, 103, 133; fixation 75–76;
   self-preservation 3, 5, 8, 24–28, 39–43,
   45, 49; sexual 77, 102; scopic 112–113

Edelman, G. 12–13, 16, 41, 83
enjoyment 27, 122; *see also* jouissance
erogenous areas or erogenous parts 77, 102

equivocal 35, 79
exogenist 70
experience 3–7, 9, 11–13, 19, 21, 27, 31, 36–37, 39, 52, 56, 67, 71, 75, 77, 79, 81, 84, 86, 94–95, 105, 111, 113, 119–121, 133–136; traumatic 21, 28, 94
extimacy 124

facilitations 71–72
failed operations 77–79; *see also* equivocal; forgot; mistake
fiction 111, 120–122; *see also* homo: narrans
Fliess, W. 73–76
forgot 73, 132; *see also* equivocal; mistake
formation: symptom 81, 87–88

Gage, P. 9
Goodman, N. 85
group 17, 80

Han, B. 35, 37–39, 114, 136–137
hippocampus 10, 39–40, 44
homeostasis 3–4, 10–13, 28, 45
homo: sapiens 112; narrans 112
hopeful expectation 49
hypothalamus 30
hysteria 8, 21, 23–24, 67–68, 136

id 39, 89–90; *see also* awareness: affective
ideal 30–31, 38, 64, 68, 73, 86, 89, 92–93, 105, 118, 121, 134
identification 16–18, 31, 47, 77, 105, 109, 121–122, 143
impulse 25, 40, 44–46, 71
imputations to psychoanalysis 45–50, 131–140
induction 85–86
inertia 9; principle of 70
infatuation 79–85
inference 40, 88
information: cognitive 6, 8, 12, 15, 31, 33–39, 53–55, 115
instinct 3–4, 10, 24, 26, 30, 40–41, 45, 92
insula 3–4, 24, 49

Jakobson, R. 20
Johnston, A. 3, 8–9, 43–45, 49
jouissance 9, 13, 17, 49, 81, 88, 90, 103–104, 108–111, 113–114, 132–133; in animals 31; and NAS-DA dopaminergic system 25–29, 108; unlimited 36,

121; *see also* body and jouissance; body: Other; enjoyment; *parlêtre*; unconscious: being that speaks

Kafka, F. 30
Kandel, E. 4–5, 7–10, 12–15, 18, 20, 29–31, 33, 41, 46–47, 49, 107, 111, 114, 135–136, 138
Kernberg, O. 31
know: psychoanalytical 25, 29, 37, 45, 65, 68–69, 78–83, 87–88, 96, 102–103, 109, 111–113, 116, 119, 121, 123, 125, 132–135, 139
knowledge: scientific 13, 17–19, 22, 24, 32, 34–37, 43–44, 46–51, 53, 55, 64, 72, 75–76, 85–86, 93, 95, 103, 105, 107–108, 110, 113, 115–117, 125, 132, 137
Kuhn, T. 1

Langaney, A. 11, 31–32, 122
language 4–5, 7, 19–20, 25, 27, 30, 36, 42, 44, 49, 66, 74–76, 80–81, 104, 108, 111, 119, 124, 134–135
Letter 52 6, 16, 75–76, 97
libido: neurochemistry of the 40–43; *see also* dopamine, instinct
lo: the term 90–91
localization 15, 19, 68, 70, 76, 89; *see also* anatomical-clinical method; functional method; unconscious: neuroanatomy theory of; unconscious: neurodynamic theory of

Magistretti, P. 3–7, 9–11, 41, 48–49
mania 29
market 35, 38, 51, 55, 80, 115–117; *see also* capitalism
master: desire to be the 48, 73, 79–80, 117; discourse of the 1, 48, 79–80; of himself 38, 124; identifications of the 18, 47; signifier 122
maturity 18
meaning 3, 7, 15, 17, 30–31, 36, 40, 44–45, 66, 72, 75, 79–80, 86, 88–89, 91–92, 111, 113, 125–126, 132, 134; *see also* sense; significance; signification; signified
memory: declarative 10–11; non-declarative 13, 15, 19, 39, 45; procedural 9–11; recovery of 3–8, 37, 39, 71–72, 75
mental health 43, 51, 53–54, 89, 93, 103, 105, 116, 125; *see also* ideal

method 8, 11, 14, 17, 29, 31, 33, 40, 46–49,
86, 96, 110, 118–119, 131–134, 136,
138; anatomical-clinical 15, 19–22, 24,
47–48, 95; *see also* theory: functional;
theory: dynamic
microworlds 80–81; *see also* specialty
mind 4, 8, 12–19, 29, 33–34, 40, 46,
50–52, 54, 56, 74, 80, 105, 107–109,
138; *see also* information: cognitive;
unconscious: mental
mistake 77–79, 84, 89
motor activation 25–28; *see also* impulse

negative therapeutic reaction 93
neurons 4–6, 9, 16, 18, 29–31, 54, 68,
70–72, 74, 111, 118, 122, 124; omega
72; phi 70–72; psi 70–72
neuroplasticity 3–6, 9, 36
neuropsychoanalysis 7, 14–15, 20–22, 24,
29, 43–45, 69, 102–103, 108, 111, 123
Neuroscientificism 118–120
nominalism 32

object: cause of desire 26, 29, 35, 65,
72–73, 80, 84, 87, 89–90, 113, 115, 118,
125, 132, 135; consumption 37, 55, 137;
evident, specific and unified 15, 17–18,
26–29, 39–41, 44–45, 49, 54–54, 109–
110, 113, 116–117, 123, 136–139
orgasm 29, 31
Other 11, 19–20, 35–38, 90–91, 113–114,
124; big or symbolic 17, 45, 73; social
3, 8

Panksepp, J. 26–28, 32, 48
paradox 38, 44, 91, 93, 107–108, 123, 137
paralysis 22–24, 67–70, 76
parlêtre 8, 104–105
passage to the act 18
perception 5–7, 39, 47, 71–72, 75, 86;
speech 30; unconscious 16–18
pharmacology 54, 114–116
phobia 81
Pinker, S. 17, 33, 46, 117
pleasure principle 41, 83, 94–95; *see also*
survival; unconscious: procedural
polymorphous perverse 76
Popper, K. 133
preconscious 8, 18–19
pressure of life 70
probability 35, 41–42
proportion 34, 37, 134; *see also*
information: cognitive

psychoanalytic punctuation 36
psychosis 36, 83; *see also* delirium
psychotherapeutic 17–18, 20, 32–33, 46–
47, 52, 54–55, 81, 84–85, 89, 114–115,
119, 122; inefficiency 131–132, 134,
138

reality: autonomous 11–13; internal 11–13;
material 55, 133, 138; physical 5–8, 19,
39, 44, 49, 69, 104, 113, 122; procedural
9–11; psychic 5–6, 8, 19, 69, 71, 76,
82, 104, 111, 113, 119, 121–122;
virtual 69
Redmond, J. 14–15, 102
register: imaginary 4, 44, 52, 74, 102, 105,
110–113, 124–125; real 29, 76, 80–81,
83, 88–89, 94, 104–105, 108–113, 115,
118, 120–122, 125, 135, 138; symbolic
9, 19–20, 29, 44–45, 74, 102, 110–111,
113, 133, 135
relationship between: cause and effect
118; conscious and unconscious 83;
Freud and Charcot 68; Freud and Fliess
74; God and brain 109; knowledge
and market 117, 137; memory and
neuroplasticity 4; neurology and
psychology 70; object and ego 44;
pharmacology and psychotherapeutics
114; psychoanalyst and school 31,
84; psychological processes and brain
organization 19–20, 32, 34, 67, 85;
psychosis and science 83; registers 45;
signifier and signified 6; soul and body
69, 107–108; unconscious and image
110; unconscious and inside 11, 28–29,
49; *see also* sexual relationship
relief 138; *see also* balance; pleasure
principle
repetition 9, 28, 81; pain 94, 113; *see also*
experience: traumatic
reward 34; system 26–29, 40, 42, 94–95
Russell, B. 90–91; *see also* lo: the term

salvation: psychoanalysis 45–50, 142
sense 79–80, 86–90, 110–111, 125; *see also*
meaning; significance; signification;
signified
sexual: excitation 70–72; excitement
89; orientation 30–31, 33–34;
relationship 31; *see also* desire: sexual
significance 12, 45, 70, 113, 115; *see also*
meaning; sense; signification; signified
significant 33, 77, 81, 85; *see also* signifier

signification 75, 120; *see also* meaning; sense; significance; signified
signified 6–7, 120; *see also* meaning; sense; significance; signification
signifier 75–76, 80, 87, 90–91, 94, 96, 102, 104, 111, 119–120, 125, 134; neuro 118, 121–124; *see also* significant
singularity 18, 50, 87, 108, 111, 135
small other 45, 74, 108
Solms, M. 10, 13–17, 19–22, 24–25, 32–33, 39–42, 44–45, 47–48, 116, 123, 132
speaker 8, 20, 30, 35, 47, 138; *see also* jouissance of the body; *parlêtre*
specialty 80–81
speech: freedom of 17; images and 15; motor theory and 30; objective 29; probability and 35; routines and 12
subject: cartesian 31; death of 29, 37–38, 72–73, 81, 116; desiring 29; neural 4–9, 11–13, 16–17, 25–26, 30–34, 37, 39, 43, 45, 48, 52, 54, 74, 83, 108–109, 111, 114, 116, 122; subversion of 124; unconscious 5, 8, 12, 17, 19–20, 23, 36, 43, 45, 47, 67, 69, 74, 76, 78–81, 84, 87–89, 93, 96, 102, 104–105, 110, 112–113, 115, 118–121
suggestion 47, 80, 91
superego 24, 90, 93
survival 10–12, 27–28, 33, 41, 49, 74, 94, 103, 133; *see also* homeostasis
symbolization 102
symptom 17–19, 24, 31, 68, 70, 75–76, 80–81, 88–89, 92–93, 104, 110, 115, 138–139
synapse 6, 27, 30–31, 71, 114, 135

talk 18, 25, 29, 36, 45, 47, 112, 115–116, 138; *see also* word
Talvitie, V. 10, 15–17, 22, 33–34, 47
thalamocortical: circuits 13
theory: functional 20–22, 24–25, 83, 95; dynamic 10, 12, 15, 22, 29, 36, 44–45, 47

no relationship: between neurology and meaning 25, 41, 47, 65–66, 70, 73, 75, 82, 87, 90, 92, 97, 102–104, 112, 118–121, 134–135; between neurology and sexuality 30–31, 34, 42, 70, 76–77; intracerebral 43
Tononi, G. 12–13, 16, 41, 83
trace 4–7, 13, 22–23, 31, 36, 45, 75, 78, 87
training: psychoanalyst's 36–37
transference 32, 37, 49, 73, 93, 96, 111, 116, 121
triple stratification 76

unconscious: being that speaks and 29–33, 76–79, 86–90, 91–97; cognitive 33–39, 52–54, 74–75, 79–85; drive and 102–103; ineffectiveness and 131–138; internal and autonomous 11–13, 71; intracerebral differentiation of 43–45; jouissance and 94–96, 103–105; latent 88; and memory 5–8, 75–76; mental 14–19; motor activation determines 25–29; neuroanatomy theory of 19–21; neurochemical basis of 40–43; neurodynamic theory of 15, 22–25; and neuroplasticity 5, 9, 71–73; the problem in localization of 85–86, 90–91; procedural 9–11; repressed 88; *see also* discourse: unconscious; information: cognitive
utilitarianism 132; *see also* psychotherapeutic inefficiency

violence 38–39; *see also* accumulation; capitalism; information: cognitive

wake up 125
Wallerstein, R. 32, 81
Wernicke, C. 19, 24, 89
word: rejection of the 7–8, 20, 28–30, 33, 39–40, 48–50, 68–69, 74, 76, 83, 87, 111, 113, 119, 125, 134–135, 139; *see also* hopeful expectation

For Product Safety Concerns and Information please contact our EU
representative GPSR@taylorandfrancis.com
Taylor & Francis Verlag GmbH, Kaufingerstraße 24, 80331 München, Germany

www.ingramcontent.com/pod-product-compliance
Lightning Source LLC
Chambersburg PA
CBHW050608280326
41932CB00016B/2964